Debt and Death in Rural India

Debt and Death in Rural India

THE PUNJAB STORY

Aman Sidhu
with
Inderjit Singh Jaijee

SAGE www.sagepublications.com
Los Angeles • London • New Delhi • Singapore • Washington DC

First published in 2011 by

 SAGE Publications India Pvt Ltd
B1/I-1 Mohan Cooperative Industrial Area
Mathura Road, New Delhi 110 044, India
www.sagepub.in

SAGE Publications Inc
2455 Teller Road
Thousand Oaks, California 91320, USA

SAGE Publications Ltd
1 Oliver's Yard, 55 City Road
London EC1Y 1SP, United Kingdom

SAGE Publications Asia-Pacific Pte Ltd
33 Pekin Street
#02-01 Far East Square, Singapore 048763

Published by Vivek Mehra for SAGE Publications India Pvt Ltd, typeset in 10.5/12.5 Minion Regular by Tantla Composition Private Limited, Chandigarh and printed at Chaman Enterprises, New Delhi.

Library of Congress Cataloging-in-Publication Data
Sidhu, Aman
    Debt and death in rural India: the Punjab story/Aman Sidhu with Inderjit Singh Jaijee.
        p. cm.
    Includes bibliographical references and index.
    1. Farmers—Suicidal behaviour—India—Punjab. 2. Farmers—India—Punjab—Economic conditions. 3. Water-supply—India—Punjab. 4. Agriculture and sate—India—Punjab. I. Jaijee, Inderjit Singh, 1931– II. Title.
    HV6545.35.S53   362.280954' 552091734—dc23   2011   2011032614

ISBN: 978-81-321-0653-1 (HB)

The SAGE Team: Gayeti Singh, Arpita Dasgupta, Anju Saxena and Deepti Saxena

Dedicated to Aman and Anahat

Photograph from L–R: Daljit (Aman's mother), Aman and Anahat

Aman Sidhu and her daughter Anahat died in a road accident in 2006. At that time, she was engaged in research for her doctorate on the subject of rural suicides in Panjab University's Department of Sociology. As a researcher for Movement Against State Repression (MASR), the doctorate was an extension of the work that she had been doing for several years among village families of southeastern Punjab, whose breadwinners had committed suicide. After her death, her research notes and data were consolidated and updated.

Inderjit Singh Jaijee
(Aman's father)

Thank you for choosing a SAGE product! If you have any comment, observation or feedback, I would like to personally hear from you. Please write to me at <u>contactceo@sagepub.in</u>

—Vivek Mehra, Managing Director and CEO,
SAGE Publications India Pvt Ltd, New Delhi

Bulk Sales

SAGE India offers special discounts for purchase of books in bulk. We also make available special imprints and excerpts from our books on demand.

For orders and enquiries, write to us at

Marketing Department
SAGE Publications India Pvt Ltd
B1/I-1, Mohan Cooperative Industrial Area
Mathura Road, Post Bag 7
New Delhi 110044, India
E-mail us at <u>marketing@sagepub.in</u>

Get to know more about SAGE, be invited to SAGE events, get on our mailing list. Write today to <u>marketing@sagepub.in</u>

This book is also available as an e-book.

Contents

List of Tables

List of Figures

List of Maps

Preface

When the research that constitutes the core of this book began, Aman Sidhu was 34 years old, married with a daughter aged 10, and recently returned from the UK where she had taken her MBA degree after completing a master's degree in Sociology from Panjab University, Chandigarh. She started researching on rural suicides in Lehra and Andana blocks of Moonak subdivision in Punjab's Sangrur district on behalf of the Movement Against State Repression (MASR), an NGO with which I have been associated for many years. In this work she was encouraged by her husband, Gurpreet Singh Sidhu, a progressive farmer.

The task was to enumerate, verify and document cases of suicide by farmers and farm labourers, beginning from 1988.

As the work progressed, she thought of doing a PhD on the subject, basing it on the fieldwork that she was already doing. The Department of Sociology, Panjab University, accepted her. For the next two years, she remained in contact with the Panchayats of all the villages in these two blocks and interviewed the families of all suicide victims that were reported to her.

Aman was also instrumental in starting the JSJ Degree College at Gurney Kalan (Moonak subdivision) and vocational centres for village girls. Aman's mother, Daljit Jaijee, IAS, was another powerful benevolent force in Lehra and Andana blocks. She set up the Rescue and Revival Mission with the aim of keeping the children of suicide victims in school. Her younger daughter, Harman Kaur Sharda, became the project director of this mission.

In 2006, a tragic road accident snatched the lives of both Aman and her daughter Anahat. Her PhD was incomplete but a substantial amount of statistical data and interviews had been compiled.

It was decided to take her work forward as a memorial to her. The present book is the result of that effort.

Aside from these personal circumstances, this book has been necessitated by our observation of the deepening immiseration of Punjab's rural communities. It was clear that the problems were not unique to Punjab alone. Ordinary farmers and farm labourers all over India are suffering and the causes are largely the same wherever one looks: deliberate discriminatory policies laid down by the central government. However, since Punjab's dependence on agriculture—and it is forced dependence—is so disproportionately great, the effects of these destructive policies are magnified here.

For decades India has relied on centralised planning. From the outset, the planners have concentrated all growth-stimulating measures on industry and the corporate sector. The role assigned to the farmers by these central planners has been to keep the urban workforce on an even keel by an ever-growing supply of cheap rice and wheat. The farmers were expected to do this without any help from the government in the form of subsidy. Land, water, farmers and farm labourers have been literally destroyed in the process. A segment of urban India has been fattened, rural India has been starved.

The documentation of suicide cases per se establishes the magnitude of rural Punjab's tragedy and this, in turn, compels a search for the reasons why so many people's lives were made unbearable. These reasons go beyond individual circumstances and lead inescapably to an indictment of policies at both the state and central government levels.

Political leaders and government bureaucrats are well aware of their role in the deaths of thousands of farmers and farm labourers. It is because of this knowledge that they stoutly resist calls to accurately establish the statistics of rural suicide deaths and more systematic and legally accepted recording of causes of death by the subdivisional authorities. Much of MASR's efforts over the past 20 years have been directed at pushing the government to acknowledge the reality of rural suicides and provide state assistance to the destitute families of the victims. We have not succeeded, despite a relentless letter campaign, court cases and every attempt to mobilise public opinion that our limited means allow. The state government that thinks nothing of spending lakhs on a one-day celebration pleads insolvency when asked to find the small sum that would buy widows and orphans daily bread. This latter situation is what the Rescue and Revival Mission addresses. This book might have been written faster if we had not been constantly searching for charitable individuals able and willing to donate for the sake of these suicide victims' families.

Suicides continue in villages throughout Punjab and indeed in villages throughout India. This work is far from finished.

20 June 2011   **Inderjit Singh Jaijee**
                                           Chandigarh

Acknowledgements

Grateful acknowledgements are due to Patricia Uberoi, formerly of Jawaharlal Nehru University; Professor J.P.S. Oberoi, formerly of Delhi School of Economics; Dr Gopal Iyer, formerly of Sociology Department; and Professor Emeritus H.S. Shergill, of Economics Department, Panjab University, for their expert input and criticism; and Dr Kumool Abbi and Professor Raj Mohini Sethi for their guidance to Aman. Dr G.S. Kalkat, Chairman, Punjab Farmers Commission's, was instrumental in keeping the focus on remedial measures, such as legislation for debt relief. Rajinder Singh Bains, advocate, ably represented the suicide victims' families before the Punjab and Haryana High Court. Justice T.S. Cheema gave us insight into the laws enacted to save farmers in the days of Sir Chhotu Ram. Film-makers Manmeet Singh, Harpreet Kaur, Rasil Basu and Anwar Jamal gave documentary shape to the human drama of the villages. Pritpal Singh Kochhar, on behalf of the Munshi Bishan Singh Kochhar Trust, mobilised added financial support to the needy families, and Harinder Singh and Kirandeep insured that these families received clothing for the children. We extend our gratitude to all those who have adopted the distressed families.

The Rescue and Revival Mission's village volunteers have given their time and knowledge of local society unstintingly. They are Choudhary Ram Diya Singh, Mohinder Singh and *Sarpanchs*— Kuldip Singh, Hardyal Singh, Jagan Singh, Surjit Singh, Harbhajan Singh Jaijee and Ajaib Singh. Cooperation from village leaders of Moonak and Lehra subdivisions also made the work of preparing this book easier.

The John F. Kennedy School of Government of Harvard University encouraged us by inviting our project director, Harman Kaur Sharda,

Aman's sister, to its annual International Bridge Builders Conference. The school also selected the work of Rescue and Revival Mission for a case study in grassroots initiative.

For editorial and statistical assistance, we are grateful to Dona Suri and Asha Yadav.

Greatest thanks are due to former presidents of India, K.R. Narayanan and A.P.J. Abdul Kalam for their interest, encouragement and positive intervention. Likewise, we thank the former speaker of the Lok Sabha, Somnath Chatterjee, who responded with concern.

1

Introduction*

Fifty years ago at the time of Independence, Punjab was an island of rural prosperity and the richest state in the country. Now it is struggling to provide bare survival to its 16 million rural citizens. For many of the state's debt-ridden farmers, the struggle for existence no longer seems worth the effort; Punjab has the highest number of rural suicides in the country. Against the state government's admission of 2,116 rural suicides since 1988, MASR already has a record of more than 750 suicides from the same period and this is for only two blocks of Moonak subdivision (Lehra and Andana). Punjab has 138 blocks. This data does not represent the complete toll; the actual number would well exceed a thousand. The causes for Punjab's farmers' distress are not difficult to identify.

MAIN FACTORS RESPONSIBLE FOR RURAL SUICIDES

Punjab Is Unjustly Deprived of Its Rightful Share of River Waters

Table 1.1 shows the allocation of river water between the states of Punjab, Haryana, Rajasthan, Jammu and Kashmir, and Delhi, as per the Eradi Tribunal Award.

---

*In May 2005, Aman Sidhu, Researcher, Movement Against State Repression (MASR), was invited to brief the Environment Advisory Group of the Punjabis in Britain, All-Party Parliamentary Group, chaired by John McDonald MP. This chapter is based on the text of her address (see Figure 1.1).

Table 1.1:   Allocation (as per the Eradi Tribunal Award)

Availability of river water	MAF*
Total availability of river waters	17.17
Punjab	5
Haryana	3.83
Rajasthan	8
J&K	0.2
Delhi	0.2

Source: Report of Eradi Tribunal.
Notes: *MAF: million acre feet.
Water below the rim stations of the Ravi and the Beas, the lowest points at which flow is recorded, amounts to an additional 0.60 MAF.

Diversion of Punjab River Waters

The diversion of Punjab river waters to other states results in:

1. Loss to Punjab
2. Shortfall in value terms
3. Land loss
4. Canal repair costs
5. Flooding caused by neglect of canals

Punjab's Rights under Riparian Law Are Violated

Those states that are vulnerable to ravages of floods have a right to their river waters. Neither the central government nor riparian states, other than Punjab, compensate Punjab farmers for flood damage. For example, following the 1988 floods, the central government gave ₹2,700 crore to Punjab for flood damage. This money was utilised for government infrastructure repair. Not a single paisa was given to the farmers.

Punjab Does Not Fully Benefit from Electricity Generated by Its Rivers

1. Punjab's share of hydro-electric power
2. Share of other states

3. Punjab's power requirement and the shortfall
4. Shortfall met by costly thermal power generation
5. Cost to Punjab

Punjab's rivers are not interstate rivers. Their water was diverted to Rajasthan and Haryana in violation of national and international riparian law through coercive agreements. Prior to the Indus Treaty, Punjab sold some water to Rajasthan under three conditions: the deal might be cancelled at any time at Punjab's convenience; it was on payment of seigneurage; and Rajasthan would permit Punjabi farmers to settle in the areas served by the canal system.

Shortly after India signed the Indus Treaty, Punjab was told that surplus water was not to flow into Pakistan and that the water of the Ravi, Beas and Sutlej available to it was far in excess of its requirement. Regarding Rajasthan, Punjab was told that its southern neighbour would never be able to utilise the 8 MAF earmarked for it, so Punjab would be given the advantage of the unused balance. The state was pushed to sign away water to Rajasthan.

Seventy-five per cent of Punjab's river water was diverted to non-riparian Haryana and Rajasthan and seigneurage—which Rajasthan was paying—was later withdrawn to legitimise Rajasthan's claim on Punjab water. As Rajasthan started drawing more and more water, Punjab was left to face water scarcity. The areas worst affected by this scarcity are those at the tail end of the canal system where supply of irrigation water has been drastically reduced. These water-scarce areas coincide with the suicide-prone areas.

The present allocation of water to Rajasthan cannot be justified under any pretext and is, in plain terms, theft from Punjab. Punjab was never compensated. The diversion of this water has caused loss to Punjab agriculture that runs into astronomical figures. Punjab is excluded from any portion of Jamuna waters. If Haryana was prepared to divert some portion of Jamuna water to Punjab, then Punjab might be disposed to trade a like portion of water from Punjab rivers to Haryana.

Tribune News Service, dated 14 January 2003, quoted the coordinator for Ministry of Environment, P.V. Sridharan as saying that out of 138 blocks of Punjab:

...divided to assess groundwater level situation in Punjab, 84 have
been declared as dark, as they have exploited their groundwater; 16 are
grey blocks and only 38 are white, leaving scope for further utilization.
We also reject any talk of interlinking of rivers as we are seeing the
effect of river waters diversion in Punjab. Punjab is on the fast track to
desertification.[1]

Vis-á-vis Rajasthan, Punjab should return to its pre-1955
position, empowering Punjab to specify the quantum of water which
is to flow to Rajasthan on payment of seigneurage. This is beneficial
to India as the same quantity of water used in Punjab would yield
much more grain than if it were supplied to distant Rajasthan. This is
due to seepage and evaporation.

As a result of depriving Punjab of its river waters, the state must
depend on costly thermal power generation and tube well irrigation.
Its water table is sinking, requiring frequent deepening of tube-wells.
Salinity of subsoil water is worsening. Punjab has taken the right step
in annulling agreements by which Punjab's river water is supplied to
non-riparian and non-basin states.

Unfavourable Price Structure for Crops and Inputs

1. Green Revolution
2. Conversion to wheat–rice rotation
3. Increase in Punjab's area under cultivation
4. Increase in Punjab's grain production
5. Punjab's share in all-India grain procurement
6. Cost of production/price of crop
7. Loss per quintal
8. Wheat
9. Paddy
10. Cost recovery = 80 per cent of investment

Punjab has suffered decades of unrealistically low government-
set Minimum Support Prices (MSP) and high input costs. Price
of grain to consumers of other states have been subsidised at the
cost of the farmers of Punjab for many years. Punjab and Haryana

---

[1]Quote taken from *Tribune News Service*, 14 January 2003.

combined, contribute approximately 67 per cent of all wheat procured in India. Overall contribution (all grains) comes to around 60 per cent. The largest producers are automatically the largest sufferers.

The remedy is to relate MSP to the market price index till such time that farmers are in a position to match heavily subsidised import price. A buffer against influx of cheap imports under the new WTO regime should be built up.

Another measure that will benefit farmers would be to fix MSP for crops selected for diversification.

Restriction on Trade

Before Partition, Punjab prospered on account of being a trade route—both by land and water—to West and Central Asia. Cities along this trade route flourished. Its five rivers flowed into the Indus carrying merchandise to the port at Karachi for onward shipment.

Both Punjab and Bengal suffered the tragedy of Partition; Bengal was not deprived of its port or the trade that flowed through it but Punjab lost a significant economic resource. Reopening the border with Pakistan would not only allow Punjab to benefit from favourable terms for agricultural products, it would also benefit Himachal Pradesh, Jammu and Kashmir and Haryana in particular and India in general. Thus, trade routes between India and Pakistan should be reopened.

Land Ceiling

Up to 1977, India had one-party rule at the Centre. Although land ceiling is a state subject, the central government through its ruling Congress party in Punjab imposed a ceiling of 17.5 standard acres.

Growth in agricultural landholding is impossible due to fragmentation and unprofitability. Holdings have declined drastically: 75 per cent of the holdings are below 2 hectares. While the government has imposed ceiling on agricultural land, it has not done so on urban property or income. This discriminates against farmers. The ceiling laws have brought down most landholdings in Punjab to between 1 and 4 acres. It appears that the government planners decided to limit upward mobility to town-people only.

The remedy is to gradually increase land ceiling, taking care to protect the individual farmers against *arthiya*/corporate farming interests. Agricultural land should be restricted to verified agriculturalists of the state only, as was the law in British India and presently being followed in Himachal Pradesh and some other states.

SUBSIDIARY OR UNDERLYING REASONS

Inadequate Availability of Credit at Affordable Interest Rate

The vast majority of farm loans (70 per cent) are advanced by non-institutional lenders at high rates of interest. Only about 30 per cent are advanced by institutional lenders (banks and co-op societies) at a regulated rate of interest. Interest rate for institutional finance has been 16 to 18 per cent; in contrast, non-institutional loans carry an interest rate of 40 to 60 per cent.

Village land is rapidly passing from the hands of the farmers to the *arthiya*s and moneylenders. Men whose forefathers tilled the same fields for generations are now forced to sell their land, often to settle 'debts' that may be illegal, entirely fraudulent or even non-existent. To rural Punjab, land is livelihood, and more than that, it is life itself. Thousands of cases can be cited from the recent past wherein debt-trapped farmers and farm labourers, reduced to penury and threatened with the loss of their land, have committed suicide. Their number is rising at an alarming rate.

The remedy is to make institutional finance available to all farmers with interest ranging between 5 and 6 per cent. Another beneficial step would be to computerise revenue records immediately. Every farmer should be issued a passbook containing complete and up-to-date revenue records of his holding. This passbook should be the enabling document for loans and land purchase or sale.

Lack of Direct Subsidies to Farmers

According to Sompal, a former Union Minister for Agriculture, 'the total volume of subsidies available to farmers in India is just

10 per cent, which is peanuts compared to what is available to the agriculturally and industrially advanced countries; 60 per cent of the farmers received no subsidies in any form'.[2] The share of agriculture in the Eighth and Ninth Five Year Plans was ₹367 crore and ₹801 crore respectively, while the same for industry, was ₹3,608 and ₹2,765 crore.[3] Every time the government declares that it intends to 'help the farmers', it gives more subsidies to industrialists producing agricultural inputs. A man whose holding is barely 2 acres is not helped by this form of subsidy. The remedy is to give subsidies directly to the farmers.

Promotion of Unsuitable Corporate Farming

Adoption of a policy favouring corporate farming would be a terrible mistake. The majority of farmers and all others dependent on the farming community, would be rapidly forced off the land. Even if agriculture flourishes under corporate management, the agriculturists themselves would be dispossessed and impoverished. When new occupations become available to people from farming backgrounds, we will see a shift away from farming and for the farmers themselves, this shift represents free choice of a more remunerative option. It will be a good thing and in no way comparable to the desperation of the dispossessed. Corporate farming also will mean outflow of money from the state and perhaps even the nation. At this stage, the impoverished farmers are not in a position to purchase land. On the other hand, the *arthiyas* and corporate sector are more likely to acquire what land the farmers have left.

Flagging Agricultural Research/Extension Effort

Before the Green Revolution of the 1960s, the practice in Punjab was to allow part of the land to 'rest' each year; a third of the land was left fallow. This land was also available for grazing and thereby enriched by the manure of the cattle. The Government of

---

[2]P.P.S. Gill, 'Terms of Trade Are against Farmers, Says Sompal', *The Tribune*, 7 Dec 2002.
[3]Sompal's figures quoted by former Punjab Finance Minister, Capt. Kanwaljit Singh in *The Tribune* (5 November 2000). Kanwaljit Singh warns of WTO dangers.

India in collaboration with the American Ford Foundation pushed the Green Revolution onto Punjab. Punjab lost many of its traditional drought and blight resistant varieties of wheat and gained a strong appetite for chemical fertilisers, pesticides and irrigation water. Other crops such as pulses and oilseeds were substantially replaced by water-guzzling paddy. The central government encouraged this by keeping a support price for wheat and rice but none for other crops. This coincided with the diversion of Punjab's river water to Rajasthan.

Higher education and research are strongly dependent on central grants. The Centre provided research money for high-yielding hybrid varieties but not enough for improving grain quality. Punjab was pushed to become the breadbasket of India and this policy helped so long as international grain prices ruled higher than domestic prices. The increase in domestic production and fall in international prices resulted in greater consumer selectivity in grain purchase. Price wise, Indian wheat is no longer competitive. It is therefore necessary to increase allocation of funds to projects aimed at improving grain quality and until this is achieved, the MSP must be retained at a remunerative level.

Reduction in Government Investment in Agriculture

Failure to Promote Industry

In order to convert Punjab into the granary of India, the Centre denied the state heavy industry. The state's vulnerability, on account of its location on the border with Pakistan, was used to justify this denial. This might have been true decades ago when military technology was primitive, but as both states acquired air capability and missiles, it ceased to be valid. Agro-industry is often held up as ideal for Punjab, but even this was not allowed except as very small-scale units. Punjab has missed out on value addition for its agricultural produce.

The remedy is to encourage all industries—especially agro-industry—for Punjab. Agro-industry should be reserved for the agriculturalists of the state. One way to do this is to follow the pattern of cooperatives. Initial capital for these projects should be loaned at reduced interest as the farming sector has no money to invest. Value-addition needs to be substantially increased.

Inefficient Grain Procurement / Storage / Transport

Table 1.2 gives figures relating to procurement, storage and transport of grains in Punjab. The following measures would be cost-effective and would save grain through reduced handling operations.

1. FCI must retain primary responsibility for grain procurement unless states are fully in position to take up the responsibility. Switch over at this stage would result in grain loss and increased distress to farmers.
2. States should enjoy the advantage of grain storage. Eighty per cent of the grain should be stored in the producer states and

Table 1.2:   Procurement, storage and transport

Procurement of wheat and paddy from Punjab 2008–09	(in lakh tonnes)
Wheat	099.39
Paddy	079.08
Total food grains	178.47
Punjab storage capacity (including open storage) 2005–06	(in lakh tonnes)
FCI	71.05
Food and Civil Supplies	07.69
Punjab Civil Supplies Corp.	20.91
Markfed	36.10
State Warehousing Corp.	36.06
Central Warehousing Corp.	07.00
Mandi Board	00.54
Agro-industries Corp.	13.10
Total storage capacity	192.45
This capacity is inclusive of open plinths	
Number of trucks registered in Punjab as on 31 March 2006	1,07,534

Source: Figures acquired from personal interview with Dr H.S. Shergill.

the remainder transferred to scarcity areas. Given modern road transport facilities, grain can be shipped to any part of India within five days.

3. The producer states should have the benefit of transportation for the sake of generating local employment to transportation companies. For more details, see pp. 109–10.

Failure to Provide Education

The areas of Punjab most affected by suicides are also the areas having the lowest literacy rates. Literacy figures for Lehra block are 29 per cent, for Andana block is 28 per cent and for Budhlada block it is 28 per cent.

Low literacy means severely restricted life opportunities. As farming income nosedives, the people of these areas are unequipped to take up any other livelihood except labour. For years, 70 per cent of central grant for education has gone to colleges and universities.

It is time to change the policy and turn that 70 per cent grant to primary education. The 2004 Union Budget provided a 2 per cent additional cess for education but that is not enough and the funds generated by this cess too are likely to go to higher education rather than where it is most needed.

Centralisation of Power

The central government is directly responsible for the pauperisation of the rural sector.

All the factors cited are beyond the control of the farmers and the state government. As Punjab has the highest involvement in agro-activity, it has suffered the most.

The remedy is to vest the state with fiscal powers comparable to their powers in pre-Partition era enabling them to take immediate corrective measures.

Lack of Crop Insurance

For the past 50 years, the government has been promising farmers crop insurance.

Till date, farmers are vulnerable to all the vagaries of nature while industry and commerce enjoy full insurance coverage. Occasionally the government doles out relief in the worst-affected areas as a 'favour'; however, the relief sums are paltry and come nowhere near the actual loss. Most of the time, this relief is merely 'announced'; farmers never receive a paisa.

A viable insurance cover must be provided immediately. Till such time as farmers have insurance, the state must increase ad hoc relief to farmers to ₹10,000 per acre for wheat and paddy and disburse it well in time.

Continuance of Caste-based Programmes

Small and marginal farmers' income has sunk far below the poverty line and many have lost what little land they had and have become landless. Instead of bringing more and more groups under the category of Scheduled Caste (SC), the government should abolish caste-based assistance and adopt economic status as the criteria for uplift programmes. All poor people, regardless of their caste, deserve such a move.

Weakness of the State Per Se

In 1966, the states of Haryana and Himachal (then backward areas) were carved out of Punjab. These states have since surged far ahead of Punjab. This is primarily due to two factors: (a) deliberate neglect of Punjab on the specious plea that the weaker states need more development and (b) uninformed, weak and self-serving leadership of the state was more concerned with survival of its government than welfare of the people.

URGENT STEPS TO HELP THOSE IN NEED

Suicide Census by Gram Sabhas (Village Councils)

In order to rescue the nation's farmers, the first thing that must be done is to understand the gravity of the situation by studying the level of rural suicides. In his most recent fourth report, *Jai Kisan: Draft National Policy for Farmers,* Professor M.S. Swaminathan has agreed with the methodology adopted by MASR. MASR finds out the details of suicide cases directly from Gram Panchayats. A sample of the form used by MASR is included in Appendix A. With the incidence of suicide rising by the day, villagers are highly concerned and are eager to cooperate in such an exercise. This data can easily be collected within two or three months.

Compensation by Central Government

The basic disbursement of relief to next of kin of suicide victims is done by the state. Although the Centre is entirely responsible for agrarian distress in the states, it does not provide relief to suicide cases. It has been seen in the southern states, as well as in this region that the state governments try to underplay the number of rural suicides cases and stall or obstruct relief packages under various pretexts due to lack of resources. Considering the large financial commitment required to tackle a crisis of this magnitude, it is the central government rather than the state governments that should provide relief.

Rescue and Revival Measures

In 1996, agricultural loans stood at ₹5,700 crore. Today it is assessed to be around 25,000 crore. Of this 25,000 crore, about 80 per cent is non-institutional and 20 per cent—amounting to ₹5,000 crore—is institutional loan. It is not asking for too much to bail out the farmers of Punjab who contribute more than 50 per cent to the national grain procurement kitty. A rejuvenated agriculture sector will generate enough money to repay the nation within a very short time.

Vis-á-vis institutional loans, the remedy is to waive all agricultural institutional loans or impose a five-year repayment moratorium, freezing the interest as well.

Vis-á-vis non-institutional loans (*arhtiya*s and moneylenders), the way forward is to set up Debt Reconciliation Boards as was done in the 1930s in Punjab.

By the beginning of the 20th century, Punjab's agricultural debt problem had become acute. Petty farmers who could not repay their loans lost their land to moneylenders. At that time, the Punjab government set up a number of inquiry commissions and brought in a number of acts to save the farmers. As a consequence of these acts, Debt Conciliation Boards were established to bring about amicable settlement between debtors and creditors. Usurious loans were outlawed and so was alienation of agricultural land. The Punjab government is now being urged to set up such debt reconciliation boards to save farmers from dispossession. It would be advisable for the Board to certify the legality of the loan and the rate of interest and insure that the debtors are informed in writing about the status of their loan regularly, as required by law. At the same time, the Board also would make sure that moneylenders report their income from money-lending in their annual income tax returns. This will help the farmers as well as the state by unearthing black money.

In the case of the tsunami disaster, nature was to blame; in the case of suicides and pauperisation of rural Punjab, the Centre's agriculture policies are directly responsible. Central response to the tsunami disaster was both prompt and creditable but the response to rural distress in Punjab is years overdue. Over the past 10 years, more farmers and farm labourers have ended their lives in Punjab than the number of people killed in India by the tsunami.

Figure 1.1: John McDonnell's letter to Aman Sidhu

John McDonnell MP
Member of Parliament for Hayes & Harlington
Constituency Office, Pump Lane
HAYES, Middlesex. UB3 3NB

Thursday, 18 May 2006

☎ - 020 8569 0010
Fax - 020 8569 0109

Aman Sidhu,
No 1501, Sector 36D,
Chandigarh,
India

Dear Aman Sidhu,

Panjabis In Britain APPG
Environment Advisory Group

The Boothroyd Room, Portcullis House, House of Commons
Wednesday 16th May 2006

**The Predicament of the Farming Community in Panjab:
Debt, Suicides, Women Rights and Landless Farmers**

I am writing to thank you for sending a DVD of your talk from Punjab for the meeting of the Panjabis in Britain All-Party Parliamentary Group and its Environment Advisory Group on 'The Predicament of the Farming Community in Panjab' held on Tuesday 16th May 2006. Your talk entitled 'Farmers in Punjab state: Debt & Suicide' was very well received. It was interesting and knowledgeable and everyone really benefited from it.

It was unfortunate that you could not be present at the meeting as there was a lot of interest in your findings. We have received a copy of your write up, which will be put on our website with your permission. The short DVD on the families of suicide victims was also quite informative.

We have had quite a lot of positive feedback from the other participants. The talks by the various speakers raised several issues of discussions and everyone felt that there was something of interest for them. The meeting was a great success and it exceeded our expectations and goals of organising it. More meetings on the environmental issues affecting Panjab are planned for the future and we hope that you will be able to participate in them. Once again, thank you for your talk at the meeting as it made it a memorable occasion.

Yours sincerely

John McDonnell MP
**Member of Parliament for Hayes & Harlington
Chair
Panjabis In Britain APPG**

website www.john-mcdonnell.net   email office@john-mcdonnell.net

Note: This letter of appreciation was sent by John McDonnell MP for Aman Sidhu. Unfortunately, it was received on the day that Aman Sidhu and her daughter died in a road accident in 2006.

2

Punjab's Land, Rivers and Climate

INTRODUCTION

At the time of Independence, the combined Punjab Province and its princely states extended over the entire lower basin of the Indus and the western side of the Jamuna basin. The total area was about 5,57,425.26 square kilometres. The rivers Jhelum, Chenab, Ravi, Beas and Sutlej—five tributaries of the mighty Indus—watered its fertile plains and eastern Punjab in particular was richly veined with canals. With Independence, came the Partition. The districts of Montgomery, Lyalpur, Multan and Jhang went to Pakistan. After independence, came a second partition when the hill districts of Chamba, Kangra, Hamirpur, Shimla and Nahan went to the new hill state of Himachal Pradesh and a third partition hived off the districts of Ambala, Sonepat, Panipat, Jind, Sirsa, Rohtak and Mahendragarh to form the state of Haryana. Today Punjab (from 29.30° N to 32.32° N latitude and 73.55° E to 76.50° E longitude) is one of the smallest states with an area of only 50,362 square kilometres. One can easily drive from the northernmost tip of the state to its southernmost tip in a single day and likewise from its easternmost to its westernmost corners.

TOPOGRAPHY

Lying between the Ravi in the northwest and the Ghaggar in the southeast, and fringed by a foothill zone and the Shivalik hills in the northeast, the state of Punjab is essentially an extensive plain. The topography of Punjab is flat. The general slope downstream averages from about 9 inches to 1 foot per mile. The state's plains slope from about 300 metres above sea level in the northeast to about 180 metres in the southwest. Its soils were laid down by the great rivers, beginning from the Pleistocene age and continuing to the present. The topography of Punjab makes it ideally suited to agriculture and less than 10 per cent of the total area is uncultivable.

From earliest times, the people of the region endeavoured to make the lands along the Indus even more productive by channelling the waters and constructing reservoirs. As engineering skills advanced, so did the system of water management, culminating in a great dam and many canal-building projects that were started by the British in the early 20th century and continue up to the present. No part of the world is more criss-crossed by canals than Punjab and the state's Irrigation Department is one of the largest in the entire country. In addition to canals, the central position of agriculture in the state's economic life has dictated the construction of a dense network of roads and railways.

REGIONS WITHIN PUNJAB

A broad overview of Punjab suggests that the state is one vast plain but viewed more closely, it is seen that Punjab has distinct regions.

The upland plains of the Upper Bari Doab, the Bist Doab and the Malwa tract, and the flood plains of the Ravi, the Beas, the Sutlej and the Ghaggar dominate Punjab's geography. But, with the Himalayas rising on the northeastern side of the state, it also gets the Shivalik hills and a rugged strip of badly eroded terrain known as the *kandi*.

DRAINAGE

The drainage system of Punjab includes three perennial rivers; one seasonal river, numerous seasonal *nadis* and hundreds of gullies. The Sutlej, the Beas and the Ravi are the perennial rivers, and the Ghaggar is a truncated, seasonal stream. Numerous *chos* (seasonal streams) come down from the foothills and gullies abound in the Shivalik hills. The discharge of the perennial rivers is subject to wide fluctuations, ranging from a flush in August to a weak current in April.

The old Punjab was a land of five rivers which renewed the fertility of the soil and provided regular water supply. It was out of these that one of the best canal systems in the world was drawn. The rivers also acted as regional divides as the various interfluvial tracts or *doabs* developed distinctive personalities through the passage of time.

Inundation or level basin system of irrigation is used extensively. The flat topography might suggest this practice but actually the constraints are such that no other system is possible. The non-runoff restriction requires either a system in which water is applied precisely according to the infiltration rate or it must be ponded.

Natural drainage ways are not readily apparent. Irrigation water passing through each *mogha* (turn-out), as well as surface runoff during the monsoon season, is utilised on that particular watercourse area as there is no provision to let it pass to an area of lower elevation. The irrigation system is so constructed that water flows from the minor canals through *moghas* supplying water to a village area.

The farmers themselves operate and maintain their distribution system. There are no headgates at the *moghas* and if a particular canal has water in it, there is water in every watercourse (*khal*) on that canal. The farmers on the watercourse use the entire flow in turns based on farm size, and each has a fixed time each week, 10 days or two weeks to use this water. This system of rotation is called *warabundi*.

The low infiltration capacities of most soils make it impossible to get sufficient water into the soil profile during one irrigation turn. The level basin system has been utilised because it exactly fits these resources and constraints. Furrow irrigation is practised with some crops and it is interesting to observe that the furrows are constructed to provide for a great deal of storage. The ponding method then, is also used with this system; the furrow is made very broad and will store sufficient water to fill the soil reservoir even though infiltration rates require that the opportunity time be much longer than the *warabundi* turn.

The Sutlej

Among the three perennial rivers of present Punjab, the Sutlej is the master stream. It has its source in one of the western Mansarovar Lakes in Tibet (China) located at 30.66°N latitude and 81.41°E longitude at elevation of 4,633 metres above mean sea level. It is an antecedent river. It existed even before the emergence of the Himalayas which it crosses during its journey to the plains.

After taking its origin, the Sutlej takes a westerly course along the slopes of Kailash mountain before entering the Indian state of Himachal Pradesh through Shipki La. Thereafter, it bends southwest and makes its way through the mountainous and hilly topography of Himachal Pradesh till it enters Punjab near Nangal. A few miles above Nangal, the river has been dammed at Bhakra in Himachal Pradesh. On entering Punjab, the river has a northwest–southeast orientation through Jaswan dun between the two ranges of Shivalik hills in Ropar district. At Ropar, it comes down from the hills and enters the plain. Here it takes a sharp westward turn to put its course through the middle of Punjab. It separates the Bist Doab in the north from Malwa in the south. At Harike it is joined by the Beas. There onwards, it adopts a southwesterly course. It marks the boundary between India and Pakistan and enters the latter at some distance west of Fazilka. Ultimately, it joins the Chenab to merge into the Indus at Mithankot in Pakistan.

Punjab owes a lot of its prosperity to the Sutlej. The Bhakra dam on the Sutlej not only provides precious hydroelectric power but also saves the state from floods. The Nangal Barrage canal originating at Nangal, the Sirhind and the Bist Doab canals starting from Ropar, the Sirhind feeder and Rajasthan canals with their source at Harike, and the Bikaner canal with its headworks at Husainiwala, all derive their water from this river.

The river has changed its course a number of times even during recorded history. Geologists believe that the Sutlej was part of the river system of the Saraswati, which flowed from the Himalayas in a southwesterly course ultimately emptying into the Rann of Kutch. A massive earthquake sometime around 2500 BC radically altered the topography of Northern India. The Saraswati disappeared and the Sutlej was pushed northward. It has been experiencing a westward shift from time to time as evidenced by the remnants of the abandoned courses. It became a part of the Indus system. The Yamuna too was a tributary of the Saraswati but the quake pushed it eastward and it joined the Ganga.

The Beas

The Beas is another important river of Punjab, rising from the southern face of Rohtang pass (4,062 metres) in Kulu district of Himachal Pradesh. After its journey through this hill state, it touches Punjab at Talwara, where a dam has been constructed across it, only to move it back to Himachal Pradesh and to enter Punjab again near Mirthal where it is joined by the Chakvi stream. In Punjab, the river joins the Sutlej at Harike after flowing for about a 150 kilometres, separating the Upper Bari Doab from the Bist Doab. The construction of the Pong dam at Talwara has drastically reduced the flood menace of the river and has permitted extension of cultivated land, especially in Dasuya, Gurdaspur and Kapurthala *tehsils*.

The Ravi

Like the Beas, the Ravi has its source in the Kulu district of Himachal Pradesh. After passing through Chamba district of the same state, it enters Punjab at the northwestern tip of Gurdaspur district. Thereafter, it acts as border between Punjab and Jammu and Kashmir for some distance. At Madhopur is located the headworks of the Upper Bari Doab canal system, taken out from this river. Some distance below, the river roughly marks the boundary between India and Pakistan before it leaves India from the west of Lopoke in Amritsar district. Throughout its course in Punjab, the Ravi flows in a comparatively narrow valley. The fluctuations in its discharge are significantly less than those of the Beas.

It is notable that all the three rivers discussed have one small perennial stream, each running parallel to and ultimately joining them: the Budha Nala in case of the Sutlej, the Kali Bein in case of the Beas and the Sakki-Kiran Nala in case of the Ravi. These streams run through the abandoned courses.

Like most of the rivers of the Indus system, the Sutlej, the Beas and the Ravi have experienced westward shifts in their courses. Farrel's Law states that the objects in motion in the northern hemisphere have a tendency to drift to their right.

The Ghaggar

The Ghaggar is at present an inland and a largely seasonal river and not a tributary of the Indus. It takes its origin from the lower Himalayas in Sirmaur district of Himachal Pradesh. After a short journey through this state, it enters Haryana near Malla, and at a short distance below it enters Punjab near Mubarakpur. It traverses through the eastern part of Patiala district and forms the southeastern boundary of the state with Haryana. It finally leaves Punjab near Sardulgarh, passes into Haryana and then moves into Rajasthan to lose itself in the sand at a place near Hanumangarh.

The Ghaggar is notorious for floods all along its course during the rainy season. It has a wide bed throughout but contains only a narrow channel of water. It is a misfit stream. It is said to have been a prominent, perennial stream with a wide drainage basin in the past, but with the process of river capture through which this river lost a number of tributaries in its upper course, it has now been reduced to a seasonal stream. The same earthquake that altered the course of the Sutlej left the Ghaggar as a truncated stream.

Seasonal Streams (*Chos*)

In addition to the four rivers described, scores of seasonal streams, locally known as *nadis* or *chos*, present another notable feature of the Punjab drainage system. These seasonal streams are seen in the area immediately south and southeast of the Shivalik hills. Most of them start from the southern slopes of the Shivaliks, cut through the foothill zone for 10–15 kilometres and soon exhaust themselves or debouch into another stream. They make wide beds which are filled with sand, stones and gravel and have small steep valley walls. They are closely spaced, 3–5 kilometres on an average, throughout the foothill zone but their spacing is the closest in Hoshiarpur district where a *cho* even after every kilometre is not an uncommon phenomenon.

These *chos* have been notorious for soil erosion. They are now being channelled and diverted by building earth embankments which are reinforced by locally available boulders. At places, they are being dammed to store water for irrigation during the dry season. *Chos* can be said to define the *kandi* area.

Gullies

The other seasonal drainage lines found in the Shivalik hills and their adjoining tracts in the foothills zone are the deep narrow gullies with broken slopes. Many of the gullies, as well as the *chos* came into existence due to the deforestation of the Shivalik hills. They are a source of floods during the rainy season. It follows that while the perennial streams of Punjab provided the base for development, the *chos* and gullies have been a bane. As in the past so in future, the progress of Punjab rests on a rational management of its water resources.

THE IRRIGATION SYSTEM

With the advent of the Green Revolution, the state has developed its water resources effectively and a mesh of irrigation canals has been laid all over (see Map 2.1). However, change in the cropping pattern—from mixed cropping to wheat–rice rotation—and intensive

Map 2.1:   Punjab: Canal network

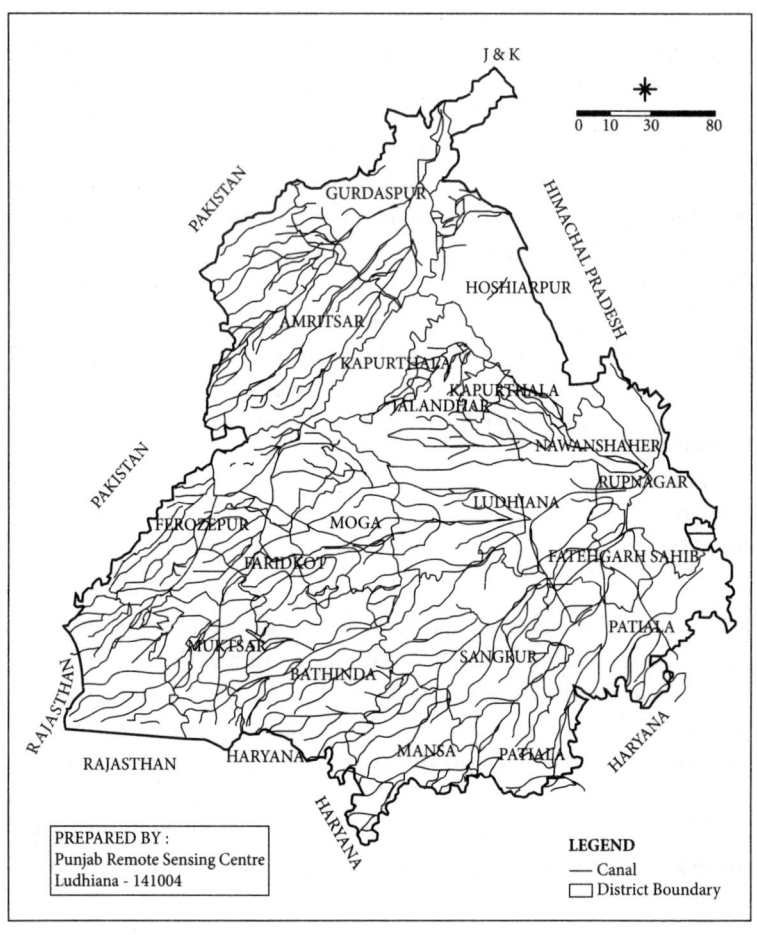

Source: Punjab Irrigation Department.

agriculture has adversely affected the state's water resources. The number of tube-wells has trebled since the 1970s. This has resulted in a decrease in the water table in certain central districts and an increase in southern districts, causing water logging and salinity problems. However, the average rate of water table decline in sweet ground water area of the state as a whole is 0.2 m/0.3 year.

Details of Capacity of Various Canals of Punjab[1]

Ground Water Status of Punjab

The available ground water is estimated at 25.34 MAF-million acre feet, where the normal requirement is 39.75 MAF.[2] This is a substantial deficit which if left uncorrected will adversely affect agriculture. The central sweet water zone, which is highly productive with a well-knit irrigation system mainly dependent upon tube-wells, presents a serious challenge because of extensive and intensive rice cultivation. In Punjab, the water table is falling at an average rate of 0.23 metres per year, during the past 15 years. The water table was at 5–6 metres in 1981, but now it shows a fall of 24–25 centimetres per annum (see Table 2.1). In only four districts of the state is the water extracted replenished by more than 50 per cent. In five districts, extraction exceeds recharge by more than 50 per cent. In two districts, recharge amounts to 20 to 50 per cent of extraction and only in one district, Amritsar, is recharge equal to extraction (see Table 2.2).

Table 2.1:   **Annual fall in the water table for districts**

Amritsar	Kapurthala	Sangrur	Sangrur	Jalandhar	Ludhiana	Faridkot	Patiala	Ropar	Gurdaspur
17	17	17	42	22	12	13	33	8	1

Source: Irrigation Department, Punjab. Also available on the website of the Punjab Soil and Water Conservation Department: http://www.dswcpunjab.gov.in/contents/punjab_waters.htm
Note: Figures in centimetre.

Table 2.2:   **District ground water recharge/extraction**

District	Ground water recharge/extraction
Gurdaspur	Recharge 20–50% of extraction
Hoshiarpur	Recharge 20–50% of extraction
Ropar	Recharge higher than 50% of extraction
Amritsar	Recharge equal to extraction (± 20%)
Kapurthala	Extraction exceeding recharge by more than 50%
Jalandhar	Extraction exceeding recharge by more than 50%

(Table 2.2 Continued)

---

[1]This section is based on data taken from Irrigation and Drainage Department, Punjab.
[2]Data taken from Punjab Agricultural University.

(Table 2.2 Continued)

District	Ground water recharge/extraction
Ludhiana	Extraction 20–50% of recharge
Patiala	Extraction exceeding recharge by more than 50%
Ferozepur	Recharge higher than 50% of extraction
Bathinda	Recharge higher than 50% of extraction
Faridkot	Recharge higher than 50% of extraction
Sangrur	Extraction exceeding recharge by more than 50%

Source: Irrigation Department, Punjab (Chandigarh).

In the southwest zone which comprises almost one-fourth of the state's cultivated area, the underground water is brackish and unfit for crops and humans. The water table in the zone which was 11 metres deep in 1981, is rising continuously at the rate of 9 centimetres per year in Mansa, 21 centimetres in Bathinda and 17 centimetres in Ferozepur. This has motivated farmers to give up cotton—once known as 'white gold'—and sow paddy instead.

CLIMATE

General

In conformity with its inland and subtropical location, Punjab is characterised by a semi-arid climate. The contrast between summer and winter is well marked and these two primary seasons are associated with two main crop seasons of Kharif and Rabi, respectively.

Seasons

The year is a cycle of five seasons: summer is from the beginning of April to the end of June, monsoon is from July to September, autumn stretches from October to mid-November, winter is from November to the beginning of February and spring is from mid-February to March end.

Summer is scorching. This is a time when rainfall is extremely rare, relative humidity is very low and dust storms are common. A very hot wind called *loo* blows in the daytime during May and June, keeping people indoors, but nights are relatively cool. The monsoons break the spell of oppressive heat. The brown landscape of summer is changed into a luxuriant green; crops flourish. Relative humidity remains high and is difficult to bear when it is calm. The post-monsoon autumn season brings a pleasant weather which is generally moderate and dry. Chill steps in soon and is at its peak during the drizzles of rainfall arising from the arrival of cyclones from the west. Frost is not uncommon but no part of Punjab receives snowfall. After the cold season, spring moderates the weather but it is only a short season. The heat picks up rapidly.

Temperature

A study of the annual temperature in Punjab confirms the continental nature of its thermal conditions. January is the coldest month, with the mean maximum temperature ranging between 20°C and 25°C and the mean minimum falling below 5°C. The mean temperature works out to be 10–15°C. The difference in the January temperature at different places in Punjab is nominal.

Temperature rises gradually during the spring but abruptly after the onset of hot season by the beginning of April. Mid-May to mid-June is the hottest part of the year with mean temperatures of about 35°C.

Subsequently, during the rainy season, temperatures fall slightly but remain stable at 25–30°C. A combination of high humidity and high temperatures makes the weather oppressive at times. The post-monsoon but pre-cold season is marked by Diwali, the main holiday of the year. This season may experience pleasant weather with the temperatures hovering around 20–25°C. Not only is there a wide range in summer and winter temperatures but a big difference is also experienced between day and night temperatures.

On an average, a day in Punjab receives 12 hours of sunshine. The span of sunshine received is the longest (14 hours) on 21 June and the shortest (10 hours) on 21 December.

There are striking regional variations in the amount of annual rainfall received in Punjab. These variations are critical to land-use pattern, agricultural productivity and population distribution. Rainfall varies from about 150 centimetres in the Shivalik hills to about a mere 25 centimetres in the southwest. The amount of rainfall declines as one moves away from the Shivalik hills, making rainfall belts parallel to the Shivalik hills. Rainfall is marked by a great seasonal concentration. About 70 per cent of the annual rainfall is received during the monsoon months of July to September. The winter showers, brought about by the western cyclones, account for 15 per cent of the total. Although small in proportion, the winter rain is critical to the success of the Rabi crops. April, May and November are the driest.

Temperature conditions remain favourable to plant growth throughout the year. The striking difference in summer and winter temperature permits cultivation of both tropical and subtropical, and even temperate crops. The abundance of sunshine is a potential source for solar energy in future.

Wind

The most common wind directions in Punjab are northwest and southeast. The former direction prevails during October to May while the latter direction is more dominant from late June to September. Winds coming from the southeast are rain bearing.

The average wind speed is 3–4 kilometres per hour. On the whole, winds are generally passive during winter, active during summer and moderate during the rainy season. The frequency of calm days is maximum during November. By comparison, wind speed is relatively high during May and June, being 5–6 kilometres in general.

Dust Storms

On an average, Punjab may experience dust storms seven days in a year. These dust storms take place usually during the summer months of May to June. They are caused either by intense low pressure at sub-regional level or by movement of dusty winds from the Rajasthan desert. Sometimes the sky remains overcast with dust for 2 or 3 days making the atmosphere suffocating. The situation improves if the dust storm is followed by rain. The frequency and intensity of these storms is higher in southwest Punjab.

In brief, climate is a mixed bag for Punjab's development, especially with respect to agriculture. While favourable thermal conditions permit plant growth throughout the year, an inadequate, seasonally concentrated and highly variable rainfall makes irrigation indispensable.

3

Cultural Profile of Moonak Subdivision

Sangrur district—of which Moonak is one of the five *tehsils* (subdivisions)—has no heavy industry and comparatively little light or small-scale industry. Agriculture dominates the economy of Sangrur. It is also among the districts of Punjab where Sikhs are in a majority; Hindus are in a majority in only four districts in Punjab: Gurdaspur, Jalandhar, Hoshiarpur and Nawanshahr.

Given these circumstances, it is no surprise that castes traditionally engaged in agriculture predominate in Sangrur district and in the Moonak–Lehra subdivision. Jats, Saini and Kamboj would account for roughly half the population, Mazhabhi Sikhs and Rajputs make up another 30 per cent, leaving all remaining castes (Brahmin, Tarkhan, Khatri, Bania, among others) to constitute the remaining 20 per cent.

CASTES

Conventional notions of caste do not apply in Punjab. In the traditional view (*varna*) Brahmins are at the top of the social hierarchy, followed by Kshatriyas, Vaishas, Sudras and finally the *Panchama*. In Punjab, the social hierarchy rests on wealth and power, both physical and political, rather than on the concepts of knowledge or ritual purity. Jats—the major land-owning group—hold the dominant position, although as per *varna* hierarchy the Jats are Sudras. Many castes are based on occupation and are named after these (e.g. Tarkhan, Sonar).

Punjab lies in the path of anyone—invading army or migrating tribe—headed southward from the Khyber Pass or eastward across the deserts of Baluchistan and Sindh. Throughout history the region has been under constant threat and has rarely enjoyed peace except for more than a few years at a stretch. In a situation where those at the top of the hierarchy today may be trampled underfoot tomorrow, as well as frequent upheavals and dislocations within, a rigid framework—social or economic—cannot gain much foothold. Islam, with its ideology of equality and universal brotherhood, exerted a strong influence on Punjab and this too worked against caste. Above all, the Sikh gurus were outspoken in their opposition to caste. They actively worked against it and made *sangat* (congregation) and *pangat* (community dining) the basis of worship.

Despite all this, it cannot be said that caste has been absent in Punjab; however, relative caste positions in Punjab have been more fluid than in other parts of India. In the past 50 years, caste has also been 'modernised' to a great extent. Today, economic, social and circumstantial factors are much more likely to influence an individual's choice of occupation than caste. Caste does not restrict an individual's associations or movement. Certainly the respect an individual enjoys in society is not dependent on caste. Caste is now a consideration only in the choice of a spouse and even in this area people are gradually becoming more flexible.

Jats constitute approximately 20 per cent of the population of Punjab and up until the end of the last century, owned about 60 per cent of agricultural lands. However, this percentage is changing as more and more hypothecated acres pass into the hands of non-Jat creditors. The Kamboj, Sainis and Rajputs are also agricultural castes but number far less than the Jats and to that extent they are less influential.

M.S. Dhami, who has spent years analysing population trends, estimates that among the Hindus, roughly 39 per cent belong to the higher castes (as against 4 per cent 'high-caste' Sikhs).[1] Among the agricultural castes, Hindus account for about 12 per cent, while the Sikhs constitute 64 per cent. Sikh artisan castes total about 12 per cent

---

[1]M.S. Dhami, 'Caste, Class and Politics in the Rural Punjab: A Study of Two Villages in Sangrur District', *Journal of Politics* 5(2), July–December 1981, pp. 93–118.

Table 3.1:   Religious communities in Punjab

Religious communities	Percentage	Persons	Males	Females
Sikh	60	1,45,92,387	76,92,776	68,99,611
Hindu	37	89,97,942	48,74,765	41,23,177
Muslim	1.6	3,82,045	2,13,023	1,69,022
Christian	1.2	2,92,800	1,54,673	1,38,127
Buddhist	0.17	41,487	22,171	19,316
Jain	0.16	39,276	18,605	20,671
Other religious communities	0.35	8,594	4,655	3,939
Total		2,43,58,999	1,29,85,045	1,13,73,954

Source: 2001 Census of India.

and the Scheduled Castes (SCs) are divided into about 40 per cent Sikhs. For a break-up of the various religious communities found in Punjab, refer to Table 3.1.

M.S. Dhami made a detailed interpretation of the 1981 Census. The 1981 Census found the population of the Punjab to be 16.8 million. Of the total Hindu population in 1981, 48 per cent lived in urban areas and 52 per cent in rural areas. The corresponding figures for Sikhs were 15 per cent and 85 per cent, respectively. Nearly two-thirds of the total Sikh population belonged to the farming castes.

In his study, Dhami found that the Sikh population had been steadily growing in the state in the 1970s; however, in the 1980s the rate of growth of Sikh population slowed down and the trend changed between 1991 and 2001.

In 1971, the Sikh population was 60.25 per cent, in 1981 it was 60.75 per cent and in 1991 it rose to 62.95 per cent. The 2001 figure was 60 per cent. There are many factors that possibly contribute to the nearly 3 per cent decline in Sikh population. This includes migration, declining fertility on account of advance in education and urbanisation, decline in the number of women (skewed gender ratio[2]) and rural suicides.

[2]Female foeticide and the Jats: Guru Gobind Singh interdicted two practices: *nadi-maar* and *kudi-maar*, i.e., consumption of intoxicants and female foeticide. This is recorded in the credo of the Sikhs or *Rehat Maryada*.

Dr Kirpal Singh has observed that in the early 19th century, the Grewals (people belonging to the Grewal *gotra* of the Jat caste) were notorious for female foeticide. The Grewals typically were small farmers and they were among the first of the Jats to get educated. Of the few daughters allowed to survive, some were sent to school and some of these educated girls got employment. This change came about in the 1930s. Dowry demand declined on account of educational status and earning capacity. Rapidly, the Grewal attitude towards their daughters changed and a situation of threat was transformed into a situation of nurturing.[3]

Today, vis-à-vis the treatment of their daughters, the Grewals are exemplary followers of the commands of Guru Gobind Singh. But it was not religious sentiment that turned the situation around; it was economics.

Another example comes from the 19th century. Maharaja Ranjit Singh's mother was the daughter of Raja Gajpat Singh of Jind. As soon as she was born she was taken out and buried alive. Bhai Guddur Singh of Bagrian, a respected Sikh leader, scolded Gajpat Singh for violating Sikh *maryada* and she was exhumed in time to save her life. She was named Raj Kaur and grew up to become a shrewd and politically astute woman. Her son founded a dynasty whose territory stretched from the Sutlej, beyond the Khyber Pass and to the border of Tibet.

Today female foeticide continues. Indeed, where economic circumstances become increasingly strained, the impetus to commit this heinous crime grows stronger. Sadly, women are often the main movers behind the decision to eliminate a female foetus, either due to economic insecurity or emotional insecurity, or both.

For more than 30 years now, Punjab has been an 'importer' of manpower for its agricultural production. These labourers are mainly drawn from eastern Uttar Pradesh and Bihar and this door opened for them when Punjab farmers began to sow paddy on a large scale.

So long as Punjab's main crop was wheat, labour needs were met locally. The exception was the Rajasthanis. When drought

---

[3]These observations on the Grewal community have been taken from a lecture delivered by Dr Kirpal Singh (Professor and Head [retired], Punjab Historical Studies Department, Punjabi University, Patiala), at the Institute of Sikh Studies.

afflicted Rajasthan, people from that state would come to work in the cotton-growing areas of Punjab but this was a sporadic phenomenon. When conditions in their home villages improved they would return.

In the case of the migrant labour that flowed in from about the 1970s onwards, conditions were agricultural work was not conducive to settling down. People from eastern and central India began to settle in Punjab when they found jobs in industries—mainly in Ludhiana and Jalandhar. They worked for lower wages than the Punjabis and were too weak to strike. Unlike some states, Punjab does not insist that entrepreneurs allot any proportion of jobs for Punjabi workers. Villagers who found that they could no longer make a go of farming found that the factory doors were closed to them. Today, the 'Bihari vote' in Ludhiana is strong enough to return Members of Parliament (MP) and Members of Legislative Assembly (MLA) and cinema halls screening Bhojpuri films flourish.

Punjab is also an 'exporter' of manpower. These are Punjabi youngsters—both Hindu and Sikh—who leave for the UK, USA, Canada, Australia, Malaysia, Thailand, European countries and the Gulf States by whatever means possible. In the past decade the Sikh population of Canada has increased by 89 percent—numbering 2.38 lakhs in 2001—adding 2 per cent to the total Canadian population. The Sikh population in the UK exceeded 5 lakh in 2001.

Between 1991 and 2001, the number of Sikhs rose from 1.28 crore to 1.46 crore, while Hindus increased from 69.9 lakhs to 89.9 lakhs. The population of Sikhs in Punjab rose by 18 lakh, registering an increase of 14 per cent, whereas, the Hindus increased by 20 lakhs during the same period—a rise of 28 per cent. In other words, the Hindu growth rate was double the Sikh growth rate.

From the late 1980s and the early part of the 1990s, persecution of the Sikhs in Punjab was at its height. Every Sikh family that could manage to send sons abroad, did so. The numbers were sufficiently high to show up in the statistics. A corresponding increase was seen in the population of Sikhs in Canada, the UK, the USA and other countries. This happened to the extent that the Sikh population dwindled and the percentage share of Hindu population in Punjab rose. Also, some Hindu families who had relocated out of Punjab in the 1980s, returned.

Muslims in Punjab have increased from 2.4 lakhs in 1991 to 3.8 lakhs—an increase in percentage as well. Christians have also increased but not significantly—from 2.39 lakhs to 2.92 lakhs.

Punjabis in general and the Sikhs in particular are cursed by a desire for sons. Despite a stringent law against gender testing and the abortion of a female foetus, people still resort to this heinous practice. The gender ratio in Punjab is 897 females per 1,000.

The farming castes are predominantly Sikh while the non-landowning higher castes following non-agricultural occupations are greatly represented among Hindus. In terms of religious composition, around 91 per cent farming castes are Sikhs while nearly 9 per cent are Hindus. The Sikh farmers are predominantly Jat Sikh—82 per cent belong to this category alone. The non-landowning Hindu castes account for about 13 per cent of the total population of the Punjab. While the higher caste groups—Brahmins, Rajputs and Khatris, and Aggarwals—account for about 13 per cent of the total population among this group, 86 per cent of these are Hindus and 14 per cent are Sikhs.

Amongst the Sikhs, speaking in terms of caste is strictly unacceptable. When the Government of India Census enumerates them on the basis of caste, it is a deliberate drag toward Hinduism. De facto, Sikhs have not been totally free of caste and its practices but the concept has no place in the Sikh religion. The manifestation of caste was further reduced under the impact of militancy. Caste has been practised among the Sikhs mainly in relation to marriages but even here there are enough exceptions to show that the caste barriers are not impermeable.

Jats

The Jats constitute the single largest group in this region; they make up about one-fourth of the population. Among the Sikhs, roughly 60 per cent are Jats. The origin of the Jats is thought to be in Central Asia and it is believed that they gradually filtered down into the Indian subcontinent from about the time of Kanishka. Chroniclers began to mention them around the 6th and 7th centuries. Today Jats are numerous in Punjab—both east and west, Haryana and western UP.

Rajputs

A few Rajputs live in Moonak subdivision and in Sangrur district. They live mostly in the northern districts of Gurdaspur and Hoshiarpur and belong to both Hindu and Sikh religions. In Punjab, they belong to the category of people pushed to the hills by invaders and their position is marginal. They are predominantly farmers, but tend to shun the hard physical work involved in farming. Indebtedness is acute among the Rajputs. While land is of supreme value for Jats this is less true of the Rajputs. More than Jats, they are inclined to leave agriculture and seek other means of livelihood.

It is believed that the Rajput clans emerged from the debris of the Kshastriya dynasties, but the exact process of the transformation is obscure. Tradition has it that the *rishi*s created four Agnikula Kshastriyas—the Prahar, Sulankhi, Panwara and Chauhan. These names were unknown to the earlier Kshastriya history. From these *Agnikula*s sprang the 36 Rajput Chatris or Rajput houses of Rajputana. Historical evidence suggests that these people are the descendants of the Huns and the Takshaks. It is quite certain that the Rajputs are a far later development than the Kshastriyas. The Rajputs are divided into several categories but the people here grade themselves as Surajbansi and Chanderbansi. Besides there are a good number of Bhatias or Bhattis who are considered to be low-grade Rajputs.

Dalits

Like the castes mentioned, Dalits—people belonging to SCs—are mainly engaged in agriculture. Punjab cannot be understood without reference to the SCs; out of every four persons in the state, one will belong to one or another of the SCs. No other state has such a high percentage of SC population. However, 37 communities come under the term SC and up to now they have not succeeded in mobilising themselves as a single force. If they are ever able to achieve this, they will transform the polity of Punjab. This may happen sooner rather than later, simply on account of the interaction of democracy and demographics. In 1991, SCs accounted for 28.3 per cent of Punjab's total population. By the 2001 census, the percentage rose to 31 per cent. However, Punjab's SCs are decidedly an economic minority: only 2.54 per cent of Dalits own any agricultural land; the vast majority is employed as agricultural labourer. Indebtedness is almost universal among agricultural labourers. Indeed, it is debt that often binds them to a particular landowner. Many castes are subsumed under the general category of SC. In Moonak and Sangrur, one generally finds people who identify themselves as Mazhabi, Adi-Dharmi, Balmiki, Ramdasia, Ravidasia, Raidasia, Kabir-panthi, Megh, Pasi, Sansi (Vimukt Jati), Bhedkut, Chamar, Jatia Chamar, Ragar, Julaha, Khatik, Kori, Koli, Bazigar, Mahesh, Sapera, Sikligar, Banglas, Dumnai, Mahasha, Dom and Sirkiband. The most numerous amongst these castes are Adi-Dharmi, Balmiki and Chamar.

The names of some of these social groupings also signify religious affiliation, like that of the Mazhabis, Adi-Dharmis, Balmikis, Ramdasias, Ravidasias, Raidasias and Kabir-panthis. Mazhabis are Sikhs. Sikhism rejects caste, nevertheless this identification remains and the Mazhabis are eligible to receive such benefits as the government claims to extend to persons belonging to SCs. Ravidasias also come within the ambit of Sikhism while Adi-Dharmis, Ramdasias and Balmikis are identified with Hindu sentiments although they strongly reject brahmanical ideology.

Chamars and Mazhabis are the most populous in this area. Most SC persons identify themselves as belonging to one or the other of these two groups.

The aim of every Dalit is to land a job in town and get out of the village and out of agricultural labour which is the traditional occupation of most Dalits. This brings not only better income and more of life's comforts but greater opportunities for their children and an enhanced sense of dignity.

4

Dispute over Punjab's River Waters

The rivers of Punjab (Jhelum, Chenab, Ravi, Beas and Sutlej) are all tributaries of the Indus and must inevitably flow from their sources in the Himalayas, down through the plains of Punjab, into Sindh and finally into the Arabian Sea. The creations of nature are not bound to abide by the creations of man, such as national boundaries. When man draws borders—'this part of the earth is mine and that part is yours'—the rivers flow from one nation to another and the issue of apportioning their waters can become contentious.

DIVIDING THE RIVERS: THE SHARES OF INDIA AND PAKISTAN AND THE SHARES OF STATES WITHIN INDIA

Immediately after the Partition, India and Pakistan spent years of intense negotiation over the details of water sharing. The Indian state of Kashmir enjoyed some benefit from the Jhelum and the Chenab but for the most part, these two rivers went to Pakistan, leaving India with only the Beas and Sutlej. The Ravi—which for some distance forms the boundary between the two Punjabs (India's and Pakistan's) is more or less equally shared by the two countries.

In 1966, another man-made boundary was created when Punjab was divided to create the new state of Haryana. This was a political division that reflected a religious divide—the Hindus dominated in the new state of Haryana and the Sikhs dominated in the remaining districts of Punjab. Water has no religion, but once again a question arose regarding how much water each state should get. As per international riparian law, only states falling within the basin of a river have a right over its waters. By that rule, Haryana has no claim because its territory lies well beyond the area drained by the Indus. Haryana's counter argument is that it is a successor state to the pre-1966 Punjab, and by virtue of that, the right over river waters devolves on the successor state. But by the same logic, Punjab, as an erstwhile part of Haryana, should have a right to a share in waters of the Yamuna. If one accepts Haryana's argument that it is entitled to Sutlej waters by the fact of being a successor state, then by the same logic, one must deny Rajasthan's claim to these waters.

The Shiromani Akali Dal (SAD) came into being in order to articulate the interests of Punjab in general and the Sikhs in particular. Retaining all the river water is clearly in the interest of Punjab which is almost wholly an agricultural state and dependent on irrigation. The Akalis have resisted allocation of river waters to other states and the allocations that have been effected have been done by the union government over the strong objections of the Akalis. This has been so from the very first decades of Independence.

The prospect of seeing Punjab's river waters 'snatched away' for Haryana took material shape in 1976 when it was proposed that a canal be built linking the Sutlej and the Yamuna; the latter is an eastward flowing river that forms the boundary between Haryana and Uttar Pradesh.

INDIRA GANDHI'S POLITICAL COMPULSIONS

In 1975, the Allahabad High Court overturned the election of Congress party leader Indira Gandhi. Rather than stepping down, she declared an Emergency which effectively made her India's dictator. From 1975 to 1977, India was under total central control.

The General Election of 1977 saw the defeat of the Congress, but within three years the victorious Janata Dal and its allies—one of which was the SAD—could not retain the confidence of the House and the government fell. Incidentally, the Janata Dal central government did not meet any of the Akali demands during its tenure (1977–79).

In the 1980 General Election campaign, the Bharatiya Janata Party (BJP) hammered the concept of *Hindutva*. Congress, under Indira Gandhi, saw its best hope of outflanking the BJP in matching their *Hindutva* card with an even more aggressive *Hindutva* strategy; Indira began to project herself as the ultimate champion of 'mainstream India'. 'Mainstream' was a codeword for Hindu India. Just as a hero needs a villain, Indira needed a threat to emphasise her ultra-Hindu identity. She found two: 'the foreign hand' (understood to mean Pakistan backed by the USA) and the Sikhs. The Sikhs were a group that was a minority in the context of India (only about 2 per cent of the population) but a majority in the context of Punjab (about 60 per cent of the population in Punjab).

It was during this period that Indira Gandhi pushed through a reallocation of Punjab's river waters. Punjab had a Congress government at that time but even the Congress Chief Minister, Darbara Singh, hesitated to assent to the new allocation, knowing the consequences that were bound to follow in the state. Indira got her way and by a notification under Section 78 of the Punjab Reorganisation Act 1966, the new allocation gave 3.5 MAF each to Punjab and Haryana, with the remaining 0.2 MAF going to Delhi. In order to help Haryana to make full use of its allocation, the construction of Sutlej–Yamuna Link (SYL) Canal was proposed. However, actual work on this project did not start until 1981.

PUNJAB RESISTS

It was in the four years preceding the actual start of work that conflict between the people of Punjab, led by the Akalis, and the union government became increasingly acute, finally erupting into bloodshed. Between August 1982 and June 1984, 410 people were killed in Punjab. Nearly 300 were killed in the first six months of 1984.

Before 1977, the SAD had been an active participant in the Janata Dal's anti-Emergency agitation and had hidden fugitive leaders of the Janata Dal in the Darbar Sahib complex. The anti-Congress spirit was still burning bright in the early 1980s. Indira's summary dismissal of the Badal government was one strong grievance. Another was the Centre's reallocation of Punjab's river waters using strong-arm methods.

As this agitation evolved, the Congress-led central government strategy was to warp the agitation out of its secular character. The tool that suited this purpose was a Sikh preacher named Jarnail Singh Bhindranwale. The Congress supported him covertly and encouraged him to continually inject a Sikh angle to the agitation, upping the ante for the Akali leadership at every step. This strategy worked and the agitation rapidly turned into a Sikh versus Indira conflict, which was very close to a Sikh versus India confrontation ('Indira is India' which was further extrapolated to 'Indira is Hindu, therefore Sikh versus Hindu'). The locus of the agitation showed a progression from the village Kapuri on the Punjab–Haryana border—where the SYL Canal was to take off—to the Darbar Sahib in Amritsar, the holiest shrine of the Sikhs.

The Darbara Singh government fell in 1983. It was replaced by an Akali government led by Parkash Singh Badal. The Akalis were negotiating with the central government at this time. The negotiations broke down and violence was on the rise in Punjab. The Akali government was dismissed. The immediate provocation for the dismissal was the killing of six Hindu bus passengers in Punjab.

TIMELINE

The following timeline makes the evolution of the water dispute clear:

1947 Soon after the Partition, India and Pakistan begin talks on the sharing of the Indus system. Negotiations continue for many years.

1953 Talks begin on a scheme that later becomes the Rajasthan Canal Project. In anticipation of the allocation of waters, work begins on the Bhakra–Nangal canal. In 1943, Kanwar Sain, the Chief Engineer of Bikaner state comes up the idea of a canal to bring water from

Punjab to his state. In 1954 the state of Rajasthan passes enabling legislation (Rajasthan Irrigation and Drainage Act 1954) and the project is inaugurated in 1958. In 1984, the name is changed to Indira Gandhi Canal. Indira Gandhi Canal is the largest canal in India (959 km long, starting from the Harike barrage in Punjab and extending beyond Jaisalmer). By 1986, work on the farthest branches of the canal was completed.

1955 The central government brings about an agreement on the allocation of the Ravi–Beas waters to Jammu and Kashmir, Patiala and East Punjab States Union (PEPSU), Punjab and Rajasthan.

Table 4.1 lists the allocations that were made of the Ravi and the Beas surplus waters, which was then estimated at 15.85 MAF after excluding the pre-Partition use of 3.13 MAF.

1956 Punjab and PEPSU merges with Punjab.

1960 With the assistance of the World Bank, the Indus Treaty is signed in 1960. The settlement could conceivably have taken the form of a joint integrated management of the entire Indus system by the two countries together or of a sharing of the waters in each of the six rivers (Indus, Jhelum, Chenab, Ravi, Beas and Sutlej) by the two countries. The state of political relations between the two countries probably ruled out the first. The second might have been very difficult and cumbersome to operate. In any case, it was not adopted.

What was agreed upon was the allocation of the three western rivers to Pakistan and the three eastern rivers to India. Under this

Table 4.1:   River water allocation

States	MAF
Jammu and Kashmir	0.15
Patiala and East Punjab States Union (PEPSU)	1.30*
Punjab	5.90*
Rajasthan	8.10

Source: Data taken from Punjab Irrigation Department, quoted by Dr G.S. Dhillon (Former Chief Engineer [Research] and Director [Irrigation and Power]) in *The Tribune*, Chandigarh, 27 May 2003.

Note: *The share of Punjab (7.2 MAF with PEPSU merged) had to be divided between Punjab and Haryana after the reorganisation of the states.

settlement, roughly 80 per cent of the waters went to Pakistan and 20 per cent was given to India.

In Pakistan it is often argued that this represented an act of generosity on its part, as the portion of territory that went to India was historically using only 8 per cent of the Indus waters. Indian sources put it at a higher figure, but it seems clear that the allocation of Indus waters to India was higher than the level of past use. This was because India, in putting forward its claims to the waters, argued for a substantial allocation to Rajasthan.

Punjab was informally assured that Rajasthan would not be able to utilise all the water allotted to it and the balance would return to Punjab. This allocation was being made to project greater utilisation of water, so that India's share vis-à-vis Pakistan could be increased. Without the Rajasthan component, the Indian share in the Indus system might have been smaller. The total availability of waters from the three eastern rivers to Punjab might have been less than it is now. Bhakra–Nangal might have required Pakistan's concurrence; it might not have been built or might have been a smaller project.

Punjab denies that Rajasthan has any right to the Ravi–Beas waters and should not have received any allocation.

Even as the negotiations were proceeding, Indian planners began to initiate action towards the eventual full utilisation of the three eastern rivers. Links between rivers to pool the waters of all three rivers were planned.

1966 In response to the Akali's long-pursued Punjabi Suba agitation, the union government divides Punjab into the new states of Punjab and Haryana.

1973 First version of the Anandpur Sahib Resolution is passed.

1976 The Ravi–Beas allocations run into difficulty. In 1976, the Government of India settles the dispute by a notification under Section 78 of the Punjab Reorganisation Act 1966, allocating 3.5 MAF to each state, with the remaining 0.2 MAF going to Delhi. In order to help Haryana to make full use of its allocation, the construction of the SYL Canal is proposed. Punjab is not happy with the decision and files a suit in the Supreme Court and Haryana too files a suit to compel Punjab to implement the decision.

1977 Jarnail Singh Bhindranwale takes over as head of the Dam Dami Taksal and launches *amrit parchar*.[1]

1978 The 18th All India Akali Conference of the SAD is held in Ludhiana (Punjab) and it adopts a softer version of the Anandpur Sahib Resolution. One of the major demands of this resolution is nullification of the allocation of river waters to non-riparian states and further reduction in the central government's control over hydroelectric installations.

1979 The Congress Party backs Sant Jarnail Singh Bhindranwale's candidates in the Shiromani Gurdwara Prabandhak Committee (SGPC) election.

1980 Punjab Assembly poll returns Congress to power under the leadership of Darbara Singh.

1981 Negotiations begin between the Akalis and the central government. The core Akali demands are socio-economic in nature and focused on the damage that centralisation is doing to Punjab's rural sector. Negotiations continue off and on for more than 30 months with the Centre—much to the intense frustration of the Akalis—adopting many types of tactics to stall resolution of the issue.

Even as the negotiations are going on, Prime Minister Indira Gandhi pushes through an agreement between the Chief Ministers of Punjab, Haryana and Rajasthan. Under that agreement the allocations to Punjab and Haryana are 4.22 MAF and 3.5 MAF respectively (out of a total availability which had been revised from 15.85 to 17.17 MAF). Punjab and Haryana withdraw their petitions from the Supreme Court. With this agreement, the obstacle to the construction of the SYL Canal is removed.

The Akalis launch an agitation against the SYL Canal. Thousands of people enthusiastically volunteer for this agitation. The agitation evolves, with each stage becoming more vociferous. The central

---

[1] A person may be a Sikh without making a full commitment to the commandments of Guru Gobind Singh. Making this commitment involves a ceremony that repeats the first ceremony conducted by Guru Gobind Singh himself in which he inducted five men to be his *Khalsa* (pure ones). He gave them water sweetened with sugar candy (*amrit*) to drink and spelled out the rules by which they should lead their lives. Before a person undergoes this ceremony, he receives instruction in the Sikh faith. This is called *amrit parchar*.

government grows highly apprehensive of the widespread disturbance in Punjab.

Jagbani editor, Lala Jagat Narain—a highly vocal proponent of the Hindu interests—is assassinated.

1982 Indira Gandhi lays the foundation stone for the SYL Canal at Kapuri on the Punjab–Haryana border.

The SAD launches an agitation against the construction of the SYL Canal called the Kapuri Morcha. Subsequently the Akalis organise a *rail roko* (railway block) and a *rasta roko* (road block) to prevent movement of wheat out of Punjab.

The SAD launches the Dharam Yudh Morcha (religious war). This agitation is based at the Darbar Sahib. Daily a *jatha*[2] of a thousand volunteers marches out to oppose the government and offer themselves for arrest. At the height of this campaign, more than 20,000 people are under detention.

Jarnail Singh Brar (Bhindranwale) moves into Guru Nanak Niwas, located at the periphery of the Darbar Sahib complex in Amritsar. It is for the first time that Bhindranwale and the SAD join hands.

The government releases all the arrested persons and resumes negotiations with SAD. Bhindranwale—who is neither an elected representative nor a party leader—is present during these discussions. An agreement appears to be imminent but at the last minute a spanner is thrown in the works by Haryana Chief Minister Bhajan Lal. No settlement is reached. The SAD frustration is intense. The Akalis announce a peaceful agitation at Delhi during the Asian Games. The police is mobilised to prevent all Sikhs from travelling to Delhi.

The SAD launches the Nahar Roko Morcha (Stop the Canal Campaign), an agitation aimed at obstructing work on the SYL.

1983 The Dharam Yudha agitation continues. Indira, under intense pressure from both the Janata Dal and the BJP, tries to 'out-Hindu' the BJP. She demonises the Akalis in particular and Sikhs in general.

The SAD launches the Rasta Roko Morcha aimed at obstructing road transport in Punjab.

Punjab DIG Avtar Singh Atwal is assassinated at the entrance to the Darbar Sahib.

---

[2]*Jatha* refers to a contingent, as in a military contingent or group of soldiers.

The SAD launches the Rail Roko Morcha, aimed at obstructing rail transportation in Punjab.

The SAD launches the Kam Roko Morcha, aimed at obstructing work on the SYL Canal.

Punjab's Congress (I) government, headed by Darbara Singh, is dismissed and replaced with President's rule, which means direct rule from New Delhi.

Bhindranwale moves into the Akal Takht.

1984 Longowal and Bhindranwale groups each publish *Vaisakhi* pamphlets accusing the other of betraying the Sikh *Panth.*

1984 Operation Bluestar: The Army enters the Darbar Sahib complex and simultaneously attacks about a dozen other historic gurdwaras in Punjab. Battle rages for two days leaving casualties running into thousands (the exact number is unknown).

Operation Woodrose: The Army combs Punjab, looking for young Sikh men who appear to be militants. Unknown numbers of Sikh men flee the state. Some enter Pakistan and others Nepal.

October 31: Indira Gandhi is assassinated by two Sikh guards deployed at her residence. Anti-Sikh riots rock Delhi and other cities.

1985 Prime Minister Rajiv Gandhi and Sant Harchand Singh Longowal sign an Accord on 24 July 1985 which includes a paragraph regarding the sharing of the Ravi–Beas waters. In pursuance of that Accord, the Ravi–Beas Tribunal (the Eradi Tribunal) is set up. The Accord stipulates completion of the SYL Canal by 1986. One clause of the Accord between Rajiv Gandhi and Sant Harchand Singh Longowal remains secret. Longowal keeps on assuring his detractors that once they learn about the secret clause, they would be satisfied. But before Longowal can disclose the secret clause, he is assassinated. K.S. Dhillon, Director General of Punjab Police, is immediately transferred out. This raises the suspicion that the central government did not want the killers to be identified.

1986–87 Work on the SYL Canal comes to a halt after 34 labourers and a superintending engineer at the site in Ropar district (Punjab) are shot dead.

1987 The Eradi Tribunal announces its award, allocating 5 MAF to Punjab and 3.83 MAF to Haryana, thus increasing the shares of both the states. It is able to do this by taking into account some additional availability of waters 'below the rim stations'.

At this time, it is found politically difficult to gazette the Award because it is unacceptable to Punjab, which is a troubled state. A reference back to the Tribunal is made raising certain issues (as is provided for under the Inter-State Water Disputes Act 1956) and the outcome in the form of a clarificatory or supplementary report is still awaited.

Punjab has all along been stoutly opposed to the construction of the SYL Canal which is meant to enable Haryana to use its share of the waters. It still remains incomplete. Haryana has been going to court over this and the Supreme Court has been asking the central government to ensure the expeditious completion of the SYL Canal.

1996 Haryana approaches the Supreme Court for directions to complete the SYL Canal.

2002 The Supreme Court decides in favour of Haryana and gives a ruling for the completion of the SYL Canal.

2004 The Supreme Court reaffirms order for the SYL Canal completion.

TERMINATION OF AGREEMENTS ACT 2004

In 2004, the Punjab Assembly unanimously passed the Termination of Water-sharing Agreements Bill and with the Governor's prompt assent, the Bill became an Act. It was in respect of that Act that a Presidential reference to the Supreme Court was made. That reference is still pending.

One Congress Chief Minister of Punjab, the late Darbara Singh, sacrificed the interests of Punjab in 1981, first by withdrawing the suit pending in the Supreme Court and later by signing an agreement on 31 December 1981, to share water with Haryana and Rajasthan just to save his chair. Another Congress Chief Minister, Captain Amarinder Singh, on 12 July 2004, got enacted the Punjab Termination of Agreements Act (PTAA) 2004 at the cost of his chair to safeguard the interests of Punjab.

THE LEGAL TANGLE

There are many missing links between 1981 and 2004, not to talk of the pre-1981 situation. Former SAD Chief Minister, Parkash Singh Badal, during his 1977 tenure, challenged the validity of Sections 78–80 of the Punjab Reorganisation Act, 1966 in the Supreme Court. The lesser known fact is that after the withdrawal of the waters case from the Supreme Court by the Darbara Singh Ministry, a farmers' organisation of Punjab filed a fresh writ petition in the Punjab and Haryana High Court challenging the validity of Sections 78–80 of the Act. In 1983, the case was abruptly transferred to the Supreme Court—for reasons best known to the powers-that-be—and has been waiting for a decision since then. It speaks volumes about the efficiency and fairness of the Indian judicial system. Even during the period when the case was pending in the Supreme Court, the Punjab and Haryana High Court on 15 January 2002, ordered the completion of the SYL Canal within a year.

Some eminent persons have argued that the agreements once entered into cannot be revoked unilaterally. They conveniently forget that Punjab was not only forced to withdraw the case from the Supreme Court in 1981 but was also pressurised to sign the agreement with Haryana and Rajasthan. It was under that agreement that Punjab—though a riparian state—was forced to take only a residual amount out of the total 17.17 MAF Ravi–Beas water (never estimated scientifically)assessed on the basis of pre-1981 availability.

Rajasthan's claim on the Punjab waters is based on the proceedings of a conference chaired by the then Central Irrigation Minister, Gulzari Lal Nanda, on 29 January 1955. This cannot be termed as an agreement. The subject file was marked as confidential and was not discussed in the State Assembly and as such it is not constitutionally valid.

Regarding cost of this water, it was decided that it should be worked out later separately as the conference was concerned only with the distribution of waters. This cost has not been worked out so far and the matter has remained dormant for the last 40 years.

5

History of Southeast Punjab

Map 5.1: Sangrur district map

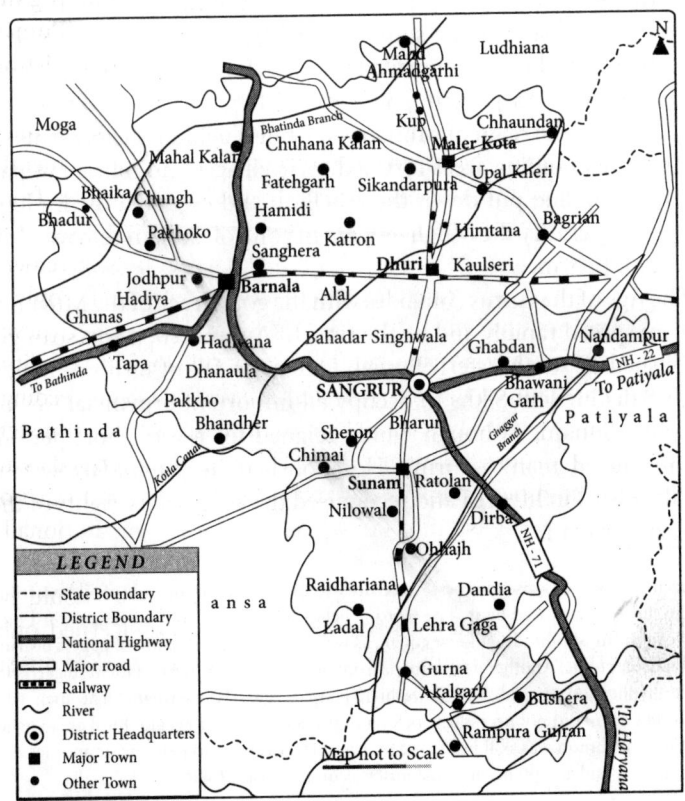

Source: Author.

BEFORE INDEPENDENCE

Trouble is Punjab's other name. Aryans, Sythinans, Greeks, Huns, Turks, Afghans, Mughals and the British—wave after wave of invasions—have swept over Punjab. The Gurus had often to fight for their survival. The Marathas carried their battles to the southern fringes of Punjab. The *misls*,[1] and later the army of Maharaja Ranjit Singh were constantly at war. Maharaja Ranjit Singh died on 20 June 1839. The British annexed the Punjab kingdom in 1849 after defeating the army of the Punjab kingdom in the First (1845) and Second (1847) Anglo-Sikh Wars.

Like most places in India, the Moonak subdivision of Sangrur district has had many rulers over the centuries. Going back to the medieval period, around AD 1100, this area looked to Bathinda—the seat of Raja Jaipal—as the source of political authority. Somewhat later, when the Khilji Sultans ruled in Delhi, Sunam was an important province. The local governor was Sher Khan and later Akhur Beg Tatak. During the reign of Muhammad-bin-Tughluq, the peasants of Sunam and adjoining areas fortified their villages and refused to pay land tax. Tughluq put down the rebellion and executed the leaders. By the 16th century the Mughals were in control. Sunam remained an important province.

The Age of the Gurus coincides with the golden age of the Mughals. Sikhism spread rapidly and by the time of Aurangzeb, it was strongly established in southeastern Punjab. After the fall of Sirhind, Banda Singh Bahadur sent Sikhs to occupy all important provincial towns, including Sunam. Although Banda reigned for a very short period, he introduced many reforms. He abolished the *zamindari* system (landlordism) in his area and recognised the proprietary rights of the tillers of the land.

---

[1]The term *misl* was first used in the 18th century to mean a unit or brigade of Sikh warriors and the territory acquired by it. The word is believed to come from a Persian word *misi* which can mean 'similitude, alike or equal', as well as 'a file' or collection of papers bearing on a particular topic. Another derivation is from *maslahai* which means a forward garrison, border fortification; armed (men), warlike (people), guards, guardians. The men who constituted a *misl* did not necessarily belong to the same caste or geographical area—what they had in common was skill in the use of arms. The decision to join one or another *misl* was voluntary and acceptance into the *misl* was on the basis of skill.

Under the Cis-Sutlej States

During the fifth Afghan invasion in 1761, Ahmed Shah Abdali defeated the Marathas at Panipat and then fell upon Ala Singh of Patiala who had sold provisions to the Marathas. Abdali sacked Barnala. Ala Singh escaped, but later on was taken prisoner. At the intercession of Shah Wali Khan, the grand *wazir* of Durrani and Najib-ud-Daula, the Rohilla Chief, his life was spared in return for a huge tribute and allegiance to the Afghans. Abdali ordered the restoration of Ala Singh's *jagirs* and gave him the title of Raja. Abdali also contributed to one of the enduring principles of the Punjabi outlook on life: *Khada-peeta laheda, baki Ahmed Shahi da* (whatever you can eat and drink is yours, the rest belongs to Ahmed Shah). Uncertainty and insecurity is a fact of life in Punjab.

The area around Sunam, including the Moonak subdivision, has its main link with the Phulkian States, Jind, Nabha and Patiala. These states were frequently at loggerheads but in the 18th century they were united in their fear and suspicion of Maharaja Ranjit Singh.

The people of Sangrur district never came under the direct control of the British government. The Federal Scheme—evolved in London at the Round Table Conference—was only concerned with British territories and did not propose to give even a little power to the people in the princely states. Whatever power fell to the share of the princely states it passed into the hands of the rulers. Even the Indian National Congress did not interfere much in the matters of the princely states. The freedom struggle of the people of southeastern Punjab was mainly the struggle against rulers who were following a repressive policy, backed by the British. The main grievance was higher *mal* (revenue) and *abaiana* (water) tax, *begar* (compulsory labour) and tenancy rights. The national movements had its impact on the people of this area and they also contributed in the main struggle for freedom from the British rule.

The Riyasti Praja Mandal

The Riyasti Praja Mandal and the Muzara Andolan centred in the areas around Sunam, Bhawanigarh and Barnala (now a part of

Sangrur district but then part of the Patiala state). These were the first grassroots political movements of modern times. For the first time, the people in the princely states openly opposed the despotic rule of the princes. Praja Mandal in the princely states and British Punjab is coterminous with the depression of the 1930s. British Punjab's agrarian situation was eased by the reforms instituted by Sir Malcolm Darling and Sir Chhottu Ram. In the princely states, reform came very slowly. Political activity was banned in the princely states so the initial stirrings for reform wore the cloak of a social movement. Even when the rulers of princely states were benign the fact was that they were monarchs and could not be held to account by the people. There were no legislative councils to articulate popular sentiment to the king. Reform could only come through agitation.

The Riyasti Praja Mandal mobilised Sikh peasants to demand redressal for their grievances, administrative reforms and the institution of responsible government. It took roots in the Singh Sabha and Akali movements in the East Punjab States. Leaders of the movement, Sardar Sewa Singh Thikriwala and Sardar Harchand Singh Jaijee, were earlier involved in the Gurdwara Sudhar Movement and in the Singh Sabha Lehar. Up to 1938, the Praja Mandal activists were affectionately called the Panth Dardis and Lok Dardis. The Muzara Movement was started by the *muzara*s (tenants) who organised them to resist paying *batai* (share of crop) to *biswedar*s (landlords).

The government's indifference to the plight of the people resulted in Jaijee and Thikriwala turning their attention to the glaring ills in public administration by the rulers. The focus was on agrarian reform, civil liberties and increased share of people of the state in the higher echelons of government. This is because, if the *mahants*[2] controlled the Sikh shrines in Punjab and non-Sikhs from outside the state held the levers of administrative control in the states, then the Sikhs were adequately represented only in the army.

Harchand Singh Jaijee and Sardar Sewa Singh Thikriwala, were co-founders of the Riyasti Praja Mandal movement and friends. During the Gurdwara Reform Movement, Sardar Harchand Singh

---

[2]*Mahant* referred to a person who was essentially the proprietor of a gurdwara, enjoying all the income of the gurdwara and conducting its activities.

Jaijee was first arrested in 1921 after the Nankana Sahib episode, and again in 1923. Later, this arrest was converted into a house arrest for five years. Sewa Singh Thikriwala was arrested in 1923, sentenced to six years in jail and detained at the Lahore Fort. In 1928, he was transferred to the jail in Patiala. Although he was a political prisoner, he was treated as a common criminal. In protest against the ill treatment meted out to him, he went on a fast. When his condition deteriorated, there was a concern in the Sikh community, both within the East Punjab States and the rest of Punjab and he had to be released.

Sardar Harchand Singh Jaijee, who was under house arrest, was exiled from the state in 1928 for his sociopolitical activities. His moveable and immovable assets were confiscated and his house at Patiala was demolished.

Harchand Singh Jaijee was the prime mover of the Riyasti Praja Mandal. This came to being in 1928 at a well-attended conference at Mansa when Sardar Thikriwala was in jail and Sardar Jaijee was in exile. Baba Kharak Singh, president of the Akali Dal, mobilised the public for the conference. The incarcerated Sardar Sewa Singh Thikriwala was made the president in absentia. At this meeting, the demand was voiced for establishment of representative government and safeguarding of people's rights. Sardar Sewa Singh Tikriwala died in jail 1934. After his death, Sardar Harchand Singh Jaijee became the de facto president of the Riyasti Praja Mandal.

During the war years, political activity ceased and the Praja Mandal split into three groups, i.e., Akali Dal, Communists and Congress. The Praja Mandal Movement became leaderless and functioned in a vacuum. The only exception was the tenant movement in the Patiala state which continued unabated throughout the war years.

Leftist Movements

The Kirti Kisan Sabha and Biswedari Movement had a strong Leftist orientation. The Praja Mandal, the Muzara Movement, the Kirti Kisan Sabha and the Biswedari Movement were so intermingled in their activities and objectives that they could be distinguished only in nomenclature. In 1942, the District Magistrate of Sunam issued notices in connection with the banning of the Communist conference at Ugrahan. The Kisan Conference was organised jointly by the Communist Party and the *muzaras* of the Punjab in general and those of *tehsils* Mansa (now in Bathinda District) and Sunam in particular, at village Rar (*Tehsil* Mansa) District Bathinda. Another Kisan conference was held in 1945 to protest the planned assault on Patiala State *muzaras* by the state and *biswedars*. In 1945, a new 21-member Muzara War Council was set up. It was more or less a Communist organisation and under its leadership armed guards were mobilised to defend gains of the movement.

AT THE DAWN OF FREEDOM

In the late 1940s, Punjab politics revolved around Master Tara Singh. When World War II ended, the British called the Round Table Conference at Shimla and there Master Tara Singh represented the Sikhs. He fiercely argued against the demand of Muhammad Ali Jinnah and the Muslim League to Partition India, forcefully reiterating that such a move could only hurt the Sikh community, which was scattered all over the province of Punjab without a majority in any district.

Master Tara Singh was especially infuriated at the prospect of Sikhs having to leave their most important and holy sites in Punjab, such as Nankana Sahib, Lahore, Rawalpindi and Faisalabad. He was one of the first leaders to recognise that it would become impossible for Sikhs to continue living in what would become the new state of Pakistan. The horrors that Tara Singh foresaw came true. Isolated and sporadic violent incidents were already erupting even as the Round Table Conference was in progress. When Sardar Patel and Nehru assured Tara Singh that India would not hesitate to fight Pakistan to save the Sikhs, he relented and accepted Partition. However, no one came to the aid of the Sikhs when the great upheaval of 1947 began. More than one million Hindus, Sikhs and Muslims were killed and at least a million families were uprooted. Tara Singh began to encourage Sikhs to leave West Punjab and find safety. He appealed to the Sikhs to help prevent violence and turned his attention to helping the refugees who were flooding in from West Punjab. The Congress forgot its assurances to the Sikh leadership about a federal constitution.

At this time Maharaja Yadvinder Singh was ruling over most of south-eastern Punjab with his capital at Patiala. He took a sympathetic attitude toward nationalist forces of the country and took a leading part in the negotiations with the British Cabinet Mission in 1946. He was the Chancellor of the Chamber of Indian Princes and so moulded the opinions of the ruling princes to bring them in line with the progressive leaders of the country and helped them achieve independence. He played a vital role in 1947 when it was feared that some of the princes might play an obstructive role. While the year 1947 brought celebrations in the south and the heartland of India, for Punjab and Bengal, there was blood and tears.

Maharaja Yadvinder Singh invited the Sikh refugees uprooted from West Punjab to settle in his state. This did not go well with the central Congress government of the time. Only the Muslims from Punjab and Bengal opted for Pakistan; the rest preferred to stay in India. The Congress leadership of the time was predominantly Hindu and from the Gangetic belt and although India was to be a secular state, 'Hindi, Hindu, Hindustan' was deeply grained in their mind. Religion, language and caste became a factor. Punjab and Bengal suffered as a consequence.

Immediately after Partition, the state of Patiala and the East Punjab States Union came into being. This became the only Sikh majority state in India. PEPSU did not last long. In 1956 it was merged into Punjab.

AFTER INDEPENDENCE

PEPSU and the Merger of PEPSU and Punjab

The Patiala and the East Punjab States Union (PEPSU) came into existence on 20 August 1948, with the integration of the princely states of Patiala, Nabha, Jind, Faridkot, Kapurthala, Kalsia, Nalagarh and Malerkotla. This Union came into being under the active guidance of Sardar Vallabha Bhai Patel who was then the Union Home Minister. Malerkotla, an independent erstwhile princely Muslim state, was declared a *tehsil* of Sangrur district.

At this time, the Jind state with minor variations was changed into Sangrur district. Some of the parts of the erstwhile Jind state were ceded to Haryana and Sunam, Bhawanigarh, Tapa and Barnala areas, which were formerly part of the erstwhile princely state of Patiala, were attached to the Sangrur district.

At the time of the merger of the princely states, Maharaja Yadvinder Singh of Patiala was made the Raj Parmukh and the Maharaja of Kapurthala was made Up-Raj Parmukh (Governor and Deputy Governor, respectively) of PEPSU for life. This assurance was short lived.

The States Reorganisation Commission, which had been appointed by the Government of India on 29 December 1953, submitted its report in 1955 and recommended the merger of the PEPSU with Punjab. The government accepted the Commission's recommendation and implemented it with effect from 1 November 1956. District Sangrur came into being with four *tehsils*, namely Malerkotla, Sangrur, Sunam and Barnala.

Punjabi Suba

Both before the Independence and afterwards, the Sikhs had demanded a state where they would be in a majority. After Partition, this demand intensified as it was felt that millions of Sikh families had suffered greatly, had been uprooted and hence they were desperate to get a secure political space for themselves.

After 1957, the union government reorganised state boundaries and created new states on linguistic basis. Through the 1950s and 1960s, Master Tara Singh led agitation after agitation demanding a Punjabi Suba. This would be a state of Punjabi speakers. But it would also be a state predominantly of Sikhs. Sikhs and Punjabi were more or less co-terminus—a fact that would provoke jugglery and deceit in the linguistic survey of 1965.

Prime Minister Jawaharlal Nehru was opposed to the creation of any state upon religious lines; however in 1966, with political pressure coming to a climax following the Indo-Pakistan War of 1965, where thousands of Sikh officers and soldiers in the Indian Army had displayed tremendous valour in defending the country, Prime Minister Indira Gandhi, the daughter and successor of Nehru, granted the demand. The state of Punjab was trifurcated. There emerged the Sikh-majority Punjab, which included the Sikh holy city of Amritsar. The Hindu-majority areas became the new state called Haryana. Also in 1966, other hill areas of united Punjab where Hindus predominated (Kangra, Shimla, Nalagarh and parts of Gurdaspur and Hoshiarpur districts) were amalgamated to form Himachal Pradesh which became a full-fledged state in 1971.

The demand for Punjabi Suba built-up throughout the early 1960s, but at the same time a tremendous energy and optimism was felt in Punjab. Relations with the Centre were still cordial and unshadowed by suspicion of any ill intention.

The Kairon Years

The early 1960s were the Kairon years—the period when the mighty Bhakra–Nangal dam and powerhouse was built, when work started in Punjab's new capital—Chandigarh—and when prestigious institutions were established.

As Minister for Rehabilitation in the chaotic days immediately after the Partition, Partap Singh Kairon handled the tough task of the resettlement of millions of refugees who had migrated from West Punjab. He proved himself not only a very able administrator but a humane one and a man of vision. Over three million people were resettled in East Punjab in new homes and often in new professions within a very short period of time.

Kairon was the last Sikh Chief Minister of joint Punjab. The political leadership that followed him largely frittered away the development he brought about.

Master Tara Singh lived to see his Punjabi Suba—a severely truncated ghost of the mighty Punjab of 1847 when the name Punjab could be applied to the whole territory north of Delhi as far as Kabul and from the fringes of Balochistan in the southwest right up to Tibet in the northeast. Master was spared worse days to come; he died on 22 November 1967.

Hindus and Sikhs in Punjab had been amicable neighbours from the beginning—indeed it was common for families to raise at least one son as a Sikh. But the long agitation for a Punjabi Suba introduced a tinge of bitterness. Many Hindus, when asked what language they spoke, told the Linguistic Survey staff that they spoke Hindi, even though they were Punjabi speakers. The idea was to keep the under-formation Punjabi Suba as small as possible and maximise the area that was to be ceded to Hindu-majority Haryana.

Punjab, Haryana and Himachal Pradesh

In 1966, Punjab was divided into three states; Punjab, Haryana and Himachal Pradesh. The smaller state Himachal Pradesh was given a separate capital and a high court. Punjab and Haryana, however, shared their high court and Chandigarh was declared a capital of both the states but Chandigarh itself was converted into a union territory.

Punjab's Distrust of the Centre

The first jolt of centre–state relation was the Punjabi Suba agitation of the 1960s. Another milestone along this road was the Anandpur Sahib Resolution of 1973 (and subsequent modifications) and by the 1980s, the rift was wide indeed. The Rajiv–Longowal Accord was an attempt to repair the damage but neither Rajiv Gandhi nor the Prime Minister who succeeded him, had the commitment and political will to honour the agreements arrived at. The Accords (and other agreements) were outside the legal framework and hence could not be enforced through the judiciary.

The Anandpur Sahib Resolution

In October 1973, the Working Committee of the Shiromani Akali Dal met at Keshgarh Sahib Gurdwara in Anandpur Sahib and asked the Government of India to hand over Chandigarh to Punjab. It also asked for the hand over of other Punjabi-speaking areas within other states and demanded an increase in the proportion of Sikhs in the Army. It criticised the 'foreign policy of India framed by the Congress Party' as 'worthless, hopeless and highly detrimental to the interests of the country, the nation and mankind at large'. Asking for a recasting of the Indian Constitution on 'real federal principles', it said:

> In this new Punjab (as in all other states) the Center's interference would be restricted to Defence, Foreign Relations, Currency and Communication, all other departments being in the jurisdiction of Punjab (and other states) which would be fully entitled to frame their own Constitution.[3]

Some of these claims were new, but their substance went back several decades to the division of India on the basis of religion in 1947. In this division the Sikhs had suffered most of all. They lost thousands of lives, thousands of acres of land they had made fertile in the 'Canal Colonies' and some sacred shrines left behind in what is now Pakistan.

Through the 1950s, the intrepid Master Tara Singh led the Akalis in the struggle for a Punjabi Suba—a separate, Punjabi-speaking and Sikh-dominated state that could compensate for the traumas of the Partition. The state was finally granted in 1966, but its extent was not what was hoped for; nor, indeed, were its powers. While the Indian Constitution hinted at autonomy for the states, the Anandpur Sahib Resolution interpreted the hint as a promise and sought to make autonomy real.

Indira Gandhi was riding high in 1973. She could boast of the victory in Bangladesh and the Centre was more powerful than ever before. Then in 1975 came the Emergency and the powers of the Centre and the Prime Minister were increased still further. The Akalis who had moved the Anandpur Sahib Resolution were put in jail. But in 1977 the Emergency was lifted, elections were called and the Congress party was comprehensively trounced.

---

[3]This is a part of the Anandpur Sahib Resolution which was later ratified and expanded at another party conference held in Ludhiana in 1978. The full text is available at www.sikhiwiki.org/index.php/Anandpur_Sahib_Resolution

Janata Dal Agitation and the Emergency

In this new political environment, the claims of the Akalis were renewed and indeed intensified. An Akali conference of October 1978 compared the 30 years of Congress rule to the Mughal days. But now that the Congress was out of power, said the Akalis, it was time for a 'progressive decentralisation of powers'.[4] The demands of the Anandpur Sahib Resolution were revived. Over 50,000 people had courted arrest during the Raj Narayan-led Janta Dal agitation.

Towards the end of 1978, the Akalis launched an agitation to fulfil the demands of the Anandpur Sahib Resolution. However, outside their fold, there were radicals who thought that nothing less than true

---

[4]Taken from the Anandpur Sahib Resolution (Resolution 1), which reads as follows:

Moved by Sardar Gurcharan Singh Tohra, President, Shiromani Gurdwara Parbandhak Committee, and endorsed by Sardar Parkash Singh Badal, Chief Minister, Punjab. The Shiromani Akali Dal realizes that India is a federal and republican geographical entity of different languages, religions and cultures. To safeguard the fundamental rights of the religious and linguistic minorities, to fulfill the demands of the democratic traditions and to pave the way for economic progress, it has become imperative that the Indian constitutional infrastructure should be given a real federal shape by redefining the Central and State relation and rights on the lines of the aforesaid principles and objectives.

The concept of total revolution given by Lok Nayak Jai Prakash Narayan is also based upon the progressive decentralization of powers. The climax of the process of centralization of powers of the states through repeated amendments of the Constitution during the Congress regime came before the countrymen in the form of the Emergency (1975), when all fundamental rights of all citizens was usurped. It was then that the programme of decentralization of powers ever advocated by Shiromani Akali Dal was openly accepted and adopted by other political parties including Janata Party, C.P.I. (M), D.M.K., etc.

Shiromani Akali Dal has ever stood firm on this principle and that is why after a very careful consideration it unanimously adopted a resolution to this effect first at the All India Akali Conference, Batala, then at Anandpur Sahib which has endorsed the principle of State autonomy in keeping with the concept of federalism.

As such, the Shiromani Akali Dal emphatically urges upon the Janata Government to take cognizance of the different linguistic and cultural sections, religious minorities as also the voice of millions of people and recast the constitutional structure of the country on real and meaningful federal principles to obviate the possibility of any danger to the unity and integrity of the country and, further, to enable the states to play a useful role for the progress and prosperity of the Indian people in their respective areas by a meaningful exercise of their powers.

independence, as in a separate 'Sikh Nation', would satisfy the *panth*. The call for Khalistan was issued from outside India by the likes of Ganga Singh Dhillon in Washington and Jagajit Singh Chauhan in London. Neither of them belonged to the Akali Dal. Sant Jarnail Singh who was propped up by the Congress repeatedly said that his demand was implementation of the Anandpur resolution and not Khalistan. However, after the attack by the Indian Army on the Golden Temple, the Sikh demand shifted from federal rights to secession from the Union of India.

The Green Revolution and Its Consequences

Consolidation of landholdings was the prerequisite of the Green Revolution. Consolidation was taken up the first time in the 1930s under the Cooperative Societies Act and later under the Punjab Consolidation of Holdings Act, 1936. However, actual consolidation of landholdings was made under the East Punjab Holdings (Consolidation and Prevention of Fragmentation) Act, 1948. In fact, by enacting this law, the consolidation of landholdings was made compulsory. It was done under the new Act in 12,628 villages. Consolidation was completed by the 1970s.

In terms of agricultural productivity, Punjab grew faster than any other state in India during the 1960s and 1970s and right up to the middle of 1980s. By the start of the 1970s, the Green Revolution had brought mechanisation, hybrid seeds, chemical fertilisers and pesticides to Punjab. Procurement was systematised under the Punjab Mandi Board. From 1961–62 to 1985–86, Punjab's annual rate of increase in food grain harvest was more than double the figure for the country as a whole. Punjab had 17,459 tractors per hundred thousand holdings while the all India figure was only 714.

The Rise of the Agrarian Class

The agrarian class of Punjab had been largely suppressed under the Mughal rule but as the Mughals weakened as a result of their own follies, and pressure built-up from the Marathas and the British, the time was ripe for a new political dispensation in Punjab. Out of the chaos of the post-Aurangzeb period arose Sikh Cis-Sutlej states and Maharaja Ranjit Singh. For the first time in centuries, the Sikh agriculturalists got a taste political dominance. Their ascendancy grew slowly throughout the British period.

Radical response to agrarian problems was much in evidence in the 1930s. The 1930s were years of economic crisis in Punjab. At the same time, marginal farmers were further marginalised. The response was the growth of the Praja Mandal in the princely states of Punjab. In rural Punjab, at the bottom of the agro-economic ladder, a shift to the Left, largely comprising small farmers began to gain momentum.

Agriculturalist assertion accelerated rapidly after Independence. Agricultural prosperity gave them a further boost and they readily plunged into the game of political parties and electoral democracy. Sikh religious institutions came under their sway. These rural-based politicians were above all pragmatists and their political style was thoroughly populist with little regard to ideology.

Punjab in the 1970s: Agrarian Crisis and the Naxalites

In the eyes of the ordinary farmer with limited land, the Shiromani Akali Dal was the party of the big landowners and the Congress was the party of the urban middle- and upper-classes. Only the parties of the Left appeared to take an interest in him and his problems.

Punjab had always been a fertile ground for the Leftist movements. Guru Nanak himself was a radical who preached the dignity of labour and urged the lowliest to claim their human birthright. He proclaimed that no man was born inferior to another and no man was unclean except by his own deeds.

In this period, Leftist movements gained ground rapidly, particularly in southeast Punjab. Naxalites gained many adherents. This alarmed both the Centre and the Parkash Singh Badal-led Akali government in Punjab. A large number of Naxalite supporters were killed in false encounters or thrown into jail and tortured. Prominent among them was Punjab Students Union leader Prithipal Singh Randhawa, who was killed on 17 July 1979, and a former Riyasti Praja Mandal Leader turned Naxalite, Hari Singh Mrigind who disappeared while in police custody. A major armed struggle developed all over the state involving 20,000 people. It was during the campaign to crush the Naxalites that the Punjab Police first resorted to 'encounter killings'.

After helping the Congress Party in imposing an unrealistic land ceiling in the agriculture sector in Punjab in the early 1970s, the Communist party withdrew from the rural areas and instead concentrated on infiltrating industry and government departments. Now, after the crisis in the agriculture sector, they are trying to make a comeback in the villages, but are finding it difficult.

One Militancy Replaces Another

Decimation of the Left in the 1970s created a political vacuum and it was in this political space that Sikh militancy expanded, drawing its followers from among the same impoverished and marginalised small farmers, artisans and landless rural people who had been attracted to the Left.

What seemed to be a conflict between a section of the Hindus and a section of the Sikhs was largely engineered by vested interests. Sikh militancy spun out of what started as a political rivalry between the Congress and the Akalis. The Indian Army attacked the Darbar Sahib in June 1984; two Sikh bodyguards shot dead Prime Minister Indira Gandhi on 31 October 1984; and immediately thereafter, rioters were let loose in Delhi and many other places with tragic consequences for thousands of innocent Sikh families. After the October 1984 riots that saw thousands of Sikhs cut down in mob fury, the Sikhs indeed felt alienated from the Government of India and a certain degree of polarisation was real but even in the worst days of turmoil in the 1980s and 1990s, most Hindus and Sikhs bore no animus for one another. Indeed, as the years of violence wore on, even ordinary people began to see through the game.

The latter part of the 1980s saw a reign of terror in the Punjab countryside: one stood equal chance of being gunned down by militants and by the police. The rule of law largely ceased to operate. Top police liked to refer to the situation as 'a war' in which they were fighting to save Punjab for India.

In time the violence abated and the Punjabis got back to trying to make some money and live well.

The Uncertain Future

Today most Punjabis would rather not be reminded of the bad old days of the recent past. However, making money and living well is exactly the issue for a large and important segment of Punjab's population. Except for a very brief period in the 1960s and 1970s, the villages have seen little development and less prosperity. If the momentum of the Green Revolution had been maintained, perhaps sufficient crumbs would have fallen to the disadvantaged sections to keep them quiet.

Instead, the situation today is that the water table is falling, excess use of fertilisers and pesticides have depleted soil heath, agriculture prices have not kept pace with input costs and indebtedness is endemic even among the better-off farmers.

Punjab, with only 1.53 per cent of the total land area of India, produces nearly 13 per cent of the total food grains—22.6 per cent of wheat and 10.8 per cent of rice—grown in the country. And yet, village economy is stagnating and urban economy has remained a backwater with none of the investment and sunrise industry influx that economic liberalisation is bringing to western and southern states of India. For skilled Punjabi workers, going abroad to the Gulf, the UK or Canada has ceased to be an option and become a duty to the family; to be carried out by hook or crook. During the militancy period, the exodus of Sikh Punjabis became a stampede due to state terror let loose on the Sikh youth resulting in countless cases of disappearances and custodial deaths.

The rural social structure has also undergone a near complete transformation over the last three or four decades. Until 1991, agricultural labourers constituted an ever-rising proportion but thereafter a decline set in and the proportion slipped from 23.82 to 16.30 per cent (see Figure 5.1). Although two-thirds of Punjab's population still lives in rural areas, only around 39 per cent of the main workers in the state are directly employed in agriculture. The all-India figure, as per the 2001 Census of India,[5] is still above 58 per cent.

---

[5]Registrar General and Census Commissioner, *Census of India 2001.*

Figure 5.1: Indian workers by sector (2001)

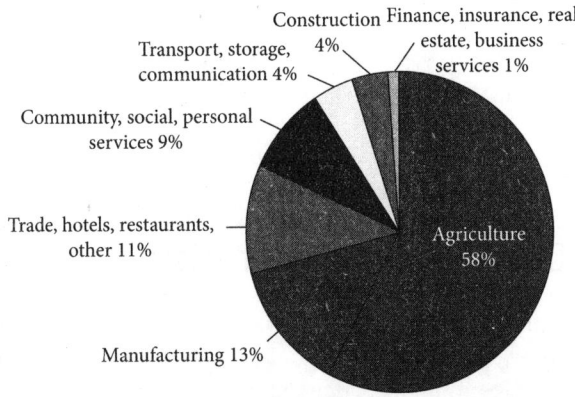

Source: Registrar General and Census Commissioner, *Census of India 2001.*

6

The Demand for Federalism

INTRODUCTION

The men who framed the Constitution of India intended the states and the central government to be bound in a federal system. However, in the very first decade following Independence, the central government began to overshadow the states and arrogate their powers. Now, more than 60 years after Independence, the Indian government shows a strong central bias.

CONSTITUTIONAL PROVISIONS VIS-À-VIS THE STATES

Representation in the Parliament can vary widely from one state to another depending on a number of factors, including demography and total land area. The Constitution puts forth the following conditions:

1. No dual citizenship.
2. The consent of a state is not required by the Parliament to alter its boundaries.
3. No state, except Kashmir, can draw up its own Constitution.
4. No state has the right to secede.
5. No division of Public Services.

Both the Union and the states are, moreover, subject to the limits imposed by the Constitution.

The Fundamental Rights cannot be violated by either the Union or the states, and thus none can be termed to be 'sovereign'. The states enjoy only relative autonomy in India. In terms of legislative, executive and judicial powers, they enjoy some freedom of action but this autonomy is limited by powers vested in the Union. There are many spheres in which the devices and the agencies of the Union can take control over matters in a state.

DISTRIBUTION OF LEGISLATIVE SUBJECTS BETWEEN THE UNION AND THE STATES

The Constitution sets out three lists by which powers are distributed between the Centre and the states.

1. The Union List: (List I) sets out areas on which the Union enjoys exclusive control. Of the total 99 subjects that are included in the Union list, some are: Banking, Coinage/Currency, Defence, Foreign Affairs, Insurance, Taxes and Union Duties.

2. The State List: (List II) mentions 69 subjects which are under the exclusive domain of the states. Some of these are: Agriculture, Fisheries, Forests, Local Government, Public Health, Public Order and Police, Sanitation, State Taxes and Duties.

3. The Concurrent List: (List III) contains 52 items, which are powers vested in the states as well as the Union. The Concurrent List is a unique feature of the Indian scheme of division of powers. The general idea underlying the Concurrent List is that there may be subjects on which the Parliament may not feel it necessary or expedient to initiate legislation in the first instance. A state may therefore make a law on a matter in that Concurrent List. However, if and when that matter assumes national importance, the Centre should

have the room to step in and enact necessary legislations in order to:

- secure uniformity in the law throughout the country, or
- guide and encourage state effort or
- provide remedies for mischief arising in the state sphere extending beyond its boundaries.

Some of the subjects under the Concurrent List are Civil Procedure, Criminal Law and Procedure, Contracts, Education, Economic and Social Planning, Marriage, Torts, Trusts, Welfare and Labour.

However, in case there is any contradiction, the union legislature will prevail over the state legislature. In case a state law has already been reserved for the consent of the President or if such an assent has already been granted, then the state law will hold. However, the Parliament can override the law through subsequent legislation.

The subject of 'education' was on the exclusive State List since the adoption of our Constitution in 1950 before a Constitutional amendment shifted it to the Concurrent List, with effect from 3 January 1977.

However, even when the Centre makes a law for the whole country on a matter in the Concurrent List, a state may also make supplementary laws on that matter to provide for special circumstances within the state. The Concurrent List thus makes the scheme of distribution of powers somewhat flexible. It permits diversity along with a unity of approach.

The residuary powers are the legislative powers that fall in none of the above categories. These powers are under the jurisdiction of the Judiciary rather than under the legislative powers of the state or the Union.

Special Situations

The legislative powers of the Parliament can be extended under special situations to include certain subjects of the State List. Here are some of the conditions under which the Parliament may extend its powers:

1. In national interest: As per Article 249 of the Constitution.
2. Proclamation of Emergency: As per Article 250 of the Constitution, the Union takes over legislative powers of the state subjects once the President declares Emergency in any state.
3. Agreement between the states: If two states agree that a given matter is related to state laws.
4. Implementation of treaties: As per Article 253, the Parliament makes laws for the implementation of treaties, even if the subject falls under the legislative power of the state, in view of the international interests of the country.
5. Failure of constitutional machinery in a state: This follows a proclamation by the President that declares the inefficiency of a state legislature. The powers of the state legislature then come directly under the jurisdiction of the union legislative. It is guided by Article 356 (1)(b).

DISTRIBUTION OF EXECUTIVE POWERS BETWEEN THE UNION AND THE STATES

The distribution of executive powers is more complicated than the distribution of legislative powers. Article 162 vests executive powers with the Union and the states on the lines of the legislative powers. Executive powers related to the laws included in the Concurrent List ordinarily remain within the state's power. However, the Union has the right to take up administration of the Union laws relative to any concurrent subject, as and when it thinks fit. The Union also has the right to mediate during any dispute between states.

The exercise of powers pertaining to the Concurrent List remains with the state, except:

1. when the Parliament vests some functions exclusively with the Union or
2. where the provisions of the Constitution itself vests some executive functions exclusively on the Union, like international agreements, irrespective of whether the subject falls in the Concurrent, Union or State List.

Separate jurisdiction of the Centre and the states is clear but a characteristic of India's federal structure in actual operation is the doctrine of 'pith and substance'. This doctrine deals with disputes between the Centre and the states, as to whether a particular law relates to a particular subject mentioned in one list or another. The courts have resolved the issue by looking at the substance of the subject rather than any incidental encroachment on it by another law.

'Entries in the legislative lists...must receive a liberal construction inspired by a broad and generous spirit and not in a narrow pedantic sense.'[1]

Even when the Parliament makes a declaration that the control of a particular industry is expedient in the public interest (Union List, No. 52), the state legislature retains its power to legislate in regard to the raw materials used in that industry.[2]

The Constitution also says that before a state legislature can pass Bills on certain matters, it must get the consent of the central executive, else the Bill will not be legally effective. For instance, a state can pass a law empowering it to acquire property bypassing the Fundamental Rights enshrined in Articles 14 and 19, but that Bill would require the President's assent before becoming a law. The state governor (appointed by the President) may refer to the President, any state legislation that endangers the high court's constitutional status or one that imposes restrictions in public interest on freedom of commerce, trade or intercourse with or within the state.

There are several instances where the central executive has had to block legislation passed by one state or another.

---

[1]Ujagar Prints vs Union of India reported in *All India Reporter* journal 1989 SC, p. 516.

[2]B. Viswanathiah & Co and Others vs State of Karnataka, 1991, Supreme Court Ruling (1) 305, SCC (3), p. 358. Delivering its judgement on 2 November 1991, the Bench ruled that:

> Legislation in regard to raw materials would be permissible under Entry 27 of List II, notwithstanding a declaration of the industry under Entry 52, to be one within the purview of parliamentary legislation. The process of manufacture or production can be legislated on by States under Entry 24 of List II so long as the industry is not a controlled industry within the meaning of Entry 7 or Entry 52 of List 1. So far as the distribution of the products of the industry is concerned, the State Legislature would be quite competent to legislate under Entry 27 of List II. However, when the industry is also a controlled industry, legislation in regard to the products of the industry would be permissible by both the Central and the State Legislatures by virtue of Entry 33 of List III.

CONSTITUTIONAL STRUCTURE

The three legislative lists in the Seventh Schedule are to be read with Parts XI and XII of the basic law, dealing with the legislative or executive relations and financial relations between the Union and the state governments. Normally, the exclusive legislative jurisdiction of the state legislatures is coterminous with their territories, legislative subjects and tax bases specified in the State List. The exclusive legislative jurisdiction of the Parliament is co-extensive with the territory of India and legislative subjects and taxes enumerated in the Union List. The Parliament also has the exclusive power to legislate on residuary matters not listed in the Concurrent List or the State List (Articles 245, 246 and 248).

The division of powers between the Parliament and the state legislatures is made on the principle of subsidiarity; i.e., state-level legislatures and the national legislature each deal with the subjects and taxes that they can best handle.

In the past 50 years, the State List has been shortened and the Concurrent List has lengthened. The Union has functionally entered into exclusive jurisdiction of the states, using its spending power via discretionary grants—as distinguished from mandatory constitutional grants—and centrally-sponsored schemes administered by the states. The formulation and monitoring of these schemes is done by ad hoc secretarial or ministerial groups chaired by the concerned union secretary or minister. Gradually constitutional amendments have altered the constitutional demarcation of the Union–State jurisdictions. Between 1950 and 2001, 27 changes were brought about by amendments: 9 in the Union List, 11 in the State List and 7 in the Concurrent List.

Union List

Of the 9 changes in the Union List, 4 have enlarged the extractive, cultural and coercive powers of the Union vis-à-vis the states. The Sixth Amendment (1956) brought taxes on sale and purchase of goods (other than newspapers) within the scope of interstate trade and commerce. The Seventh Amendment (1956) empowered the Parliament to declare a historical and cultural heritage of national importance and brought a similar power of a state legislature subject to it. The Forty-second Amendment (1976) empowered the Union to deploy armed forces or paramilitary forces of the Union in a state in aid of civil power.

State List

Barring the omissions of items from this list by amendments, the substitutions or insertions mirrored the corresponding changes in the Union List. The Seventh Amendment (1956) clarified that industries under the state's jurisdiction were exclusive of those that were declared to be under the Union jurisdiction in national interest from considerations of defence and public interest.

Concurrent List

This list has had no omissions through amendments, but it did have five additions. The Third Amendment (1954) replaced the previous entry 33 by a new one, i.e., trade and commerce whose control and regulation is warranted in public interest and welfare. The Seventh Amendment (1956) substituted the previous item of entry 42 by a new one, i.e., 'acquisition and requisitioning property' in the Concurrent List. The Forty-second Amendment (1976) added four new items to this list: administration of justice in a state and formation of lower courts (entry 11A), forests (entry 17A), education (entry 25) and population control and family planning (entry 20A). This list has seen five supplemental additions and three substitutions.

The changes have enlarged the scope of the Union and Concurrent jurisdictions and reduced the exclusive jurisdiction of the states. The Indian National Congress dominated in the first 30 years of Independence and this was the period that saw the centralisation gain, but later non-Congress regimes have not returned power to the states.

FUNCTIONAL INTERDEPENDENCE OR OVER-DEPENDENCY?

The Constitution intended a functional interdependence between the two orders of governments under normal circumstances. Article 263 of the Constitution sets forth an Inter-State Council (ISC) for harmonisation of interstate and Union–State relations and policy coordination. The composition and rules of business of this council are not specified.

The Constitution itself did not set up the ISC but merely enabled its establishment if 'it appears to the President that the public interest would be served by the establishment of a Council'.[3]

In 1990, the ISC came into being but it has made little difference. It has not superseded extra-constitutional intergovernmental forums such as the National Development Council (NDC) set up by the Nehru government in 1952 for intergovernmental approval of Five Year Plans, or several National Councils in some policy areas like local self-government, health and population.

Parliamentary enactments have also set up Zonal Councils under the States' Reorganization Act, 1956, and interstate tribunals under the Inter-State River Water Disputes Act, 1956. While Article 263 is an incontrovertible constitutional provision, governments at the Centre have more often resorted to Cabinet resolutions to set up apex forums as these allow the union government more flexibility. The Centre finds the informality of such intergovernmental forums more convenient, with the groundwork done by provisional secretarial and ministerial conferences of intergovernmental scope and by the Planning Commission officials.

Forums and tribunals such as the Zonal Councils established by the Parliament have been largely bypassed, excepting the Northeastern Council. These tribunals and judicial commissions have also tended to become mired in conflict, especially those set up to resolve river disputes. Ministers attend such council meetings with prepared speeches and there is little interaction, if any. In contrast, the ad hoc ministerial and official conferences are more open-ended. Day-to-day problems of policy and administration are deliberated on and recommendations are formulated.

---

[3]For more details, visit the ISC website at interstatecouncil.nic.in/history

The Planning Commission, for which there is no constitutional sanction at all, is the main interface between the Centre and the states. The state governments look to the Planning Commission to persuade the executive heads in New Delhi with techno-economic reason.

An exception was the Srinagar meeting of the ISC (monsoon 2003)—the first ever held outside New Delhi. A long-felt need to reform the provisions regarding the President's rule was effectively addressed at last. By consensus, members of the ISC recommended that the ruling of the Supreme Court in the case of S.R. Bommai & Others vs Union of India (1994) be written into the Constitution.[4] This landmark judgement made President's rule in a state for the first time subject to judicial review and ordered that while dismissing a state government in a 'constitutional emergency', the Union executive must not simultaneously dissolve the State Assembly until the presidential proclamation has been approved by the Parliament.

Since the 1990s, the onslaught of Centralisation has slowed slightly but the Centre's dominance is still indisputable.

All the formal or informal intergovernmental forums shun majority rule and work with consensus as determined by the Prime Minister/Union minister or Union secretary. The agenda is set by the Union in consultation with the states.

Money flows from the states to the Union by way of taxes. Two routes convey money from the Union to the states—either by discretionary grants-in-aid with or without the advice of the Planning Commission or by constitutionally mandated transfers under Article 280 of the Constitution. Many states, including Punjab, have criticised the flow of money from the Centre for being used as a tool of coercion, despite its being their rightful due, and for increasing rather than decreasing the inequality between states.

---

[4]Satish Misra, 'A Move towards Co-operative Federalism: The Srinagar Meeting Strengthens the Inter-state Relations', *The Tribune*, 6 September 2003.

> Regarding the 'emergency provisions', the meeting observed that the safeguards contained in the Bommai judgement, which have already become the law of the land, are adequate to prevent the misuse of Article 356. The Council directed that the 1994 Supreme Court judgement on in the S.R. Bommai case should be incorporated in the Constitution and asked the Centre to move expeditiously in this direction.

REFORMS

Independent India was born out of the politics of protest and this trend has, if anything, grown stronger since Independence. A heartening development of the past 25 years has been the revival of civil society institutions and growing judicial activism. However, judicial activism also carries a danger: Too much 'judicial activism' allows party leaders and leaders to feel that they can afford to make a mess and judiciary will sort it out later or that they can get away with anything if they draft the Bill cleverly enough.

India will be stronger and more democratic if it takes the political route of constitutionalism and economic development. This is also the gist of the recommendations of the Commission on Centre–State Relations (chaired by Justice R.S. Sarkaria), the Law Commissions and the National Commission for Review of the Working of the Constitution (chaired by Justice M.N. Venkatachaliah).

Rapidly the state investment in development is giving way to private sector investment and the bureaucratic License Raj state is being replaced by one where enterprise is supreme. This situation too carries the potential for distortion of democracy and demands a well thought-out partnership between the state, civil society and market forces. It is time to look at the Sarkaria and Venkatachaliah Commission reports again.

Here is a list of what the proponents of reform want:

1. Constitutional entrenchment of the National Development Council (NDC), like the ISC, as the major line organisations of intergovernmental 'executive federalism' (so called because these two apex bodies bring together executive heads of two orders of governments on a common forum for intergovernmental negotiations and policy harmonisation).

2. Streamlining of the two staff agencies of executive federalism, namely, the Finance Commission and the Planning Commission. The former serves as an autonomous advisory body under the Constitution at five-yearly intervals and the latter as a permanent secretariat of the NDC without a constitutional status. The powers of these bodies should be restricted and the powers of the states enhanced.

3. The entire spectrum of the party-system reforms recommended by the Venkatachaliah Commission Report should be implemented. These recommendations intend to bring political parties under a comprehensive law. The objective is to promote democratic and federal construction of parties with transparent and publicly accounted and audited funding.

4. Setting up of a constitutional commission on the taxation system as recommended by the Sarkaria Commission. This is necessary to deal with the constant refrain of state governments that their revenue resources are not commensurate with the heavy responsibilities the Constitution has placed on their shoulders, especially in the new context of neo-liberal economic reforms and adverse effects of globalisation on weaker sections of the population, and backward states and regions. Neither the Union nor the states can adopt an ostrich-approach to this problem. Something more than the piecemeal approach to the problem by the Raja Chelliah and Vijay Kelkar Committees on tax reforms seems to be called for.

Anandpur Sahib Resolution

Excerpt from *Anandpur Sahib: An Analysis* by Gurdarshan Singh Dhillon.[5]

Deliberate attempts to centralize power in the name of unity and integrity are incompatible with the size and diversity of the country. Such steps are bound to result in tensions and conflicts in the body politic. The demand for more state autonomy is nothing but an embittered response to such arbitrary measures as the repeated amendments to the Constitution, the imposition of Emergency and the denial of human rights and liberties to the people. The much maligned Anandpur Sahib Resolution, which has often been interpreted as a threat to the country's unity, must be viewed in this light. The Sikhs, like other ethnic groups, are struggling hard to preserve their distinctive identity and prevent the economic erosion of their state.

[5]Gurdarshan Singh Dhillon, *Anandpur Sahib: An Analysis* (Chandigarh: Ranjit Singh Riar), Preface to p. 11.

It may not be irrelevant to point out that prospects for socialism and democracy in a country, particularly of India's vastness and diversity, depend on the secular and democratic credentials of the people in power. The contention that the majority community, by virtue of its numerical superiority, might outvote the others and force on them measures, inimical to their interests and repugnant to their feelings, is not altogether baseless.

India was divided into two countries on communal lines, on the basis of a two-nation theory, the Sikhs having thrown their lot with the Hindus.

Before July 1947, the Congress had repeatedly declared that the Indian Constitution would be a federal structure with autonomous states, empowered with residuary powers. Congress leaders including Moti Lal Nehru, Mahatma Gandhi, Jawahar Lal Nehru and Maulana Azad on different occasions had interpreted Swaraj as connoting grass-roots power for the people, with greater authority vesting with the provinces.

The stand of the Congress Party was formalized in its 51st session held at Haripura in 1938 when federal system with autonomous provinces at its constituents was strongly reiterated. The Cabinet Mission also provided that only three subjects would vest with the Central Government, these being Defence, Foreign Affairs and Communications. It was in terms of this Cabinet Mission plan that the Constituent Assembly came into existence, with the full approval of the Congress.

A clear undertaking, in this regard, was given by late Jawahar Lal Nehru, while moving the executive resolution, at the opening session of the Constituent Assembly, in 1946. This Resolution envisaged 'the Indian Union as an Independent Sovereign Republic, comprising autonomous units with residuary powers, wherein the idea of social, political and economic democracy would be guaranteed to all sections of the people and adequate safeguards would be provided for minorities and backward communities and areas'. Nehru described the Resolution as 'a declaration, a pledge and an undertaking before the world, a contract of millions of Indians and, therefore, in the nature of an oath, which we mean to keep.'

Nehru at a press conference, on the eve of the All India Congress Committee meeting, at Calcutta, in July 1946, explicitly stated: 'The brave Sikhs of the Punjab are entitled to special consideration. I see nothing wrong in an area and a set up in the North, wherein the Sikhs can also experience the glow of freedom.'

The Sikhs threw their lot with the other people of India, hoping that the assurances extended to them would be fulfilled and they would be able to maintain their identity and chalk out their own development in the future. So as to become the integral part of the Indian nation, they relinquished their bargaining powers, as the third party in the political life of India. In view of their historico-political position, the British had always recognized the Sikhs as the third political entity in India, with whom questions like the transfer of power had to be discussed and finalized.

After the transfer of power, when Nehru was reminded of his declaration about the 'glow of freedom', to which the Sikhs were entitled, he rejoined that 'the circumstances have now changed'.

In 1949, when the Central Government formally elicited the views of Punjab Legislature, on the draft constitution, the Sikh representatives reiterated their stand for a federal constitution saying:

> It has been the declared policy of the Congress from the outset that India is to be the Union of autonomous states and each unit is to develop in its own way, linguistically, culturally and socially. Of course, Defence, Communications and Foreign Affairs must and should remain the central subjects. To change the basic policy now is to run counter to the oft-repeated creed of the Congress.

But in 1950 the Congress backed out from its promises to the Sikhs and its own declared objective of having a federal constitution. Instead, a constitution leaning towards a unitary form of Government was framed. In protest, the Sikh representative in the Constituent Assembly refused to sign such a constitution.

Viewed in this light, the Anandpur Sahib Resolution is the legitimate expression of the socio-political discontent of the Sikh community over the years. Delay in the creation of a Punjabi speaking state was due to the fear of Sikh domination in Punjab. The Sikhs, in turn feared the prospect of losing their distinctive identity. Unwarranted delay in the creation of Punjabi Suba led to a feeling of alienation among the Sikhs.

In 1965, a radical section of the Akali Dal, under the leadership of Master Tara Singh, raised the demand for a 'self-determined political status for the Sikhs, within the India Union.' It was argued by this section that if the Indian government could extend help to a linguistic and cultural nationality in East Pakistan, in their struggle for self-determination against the oppressive policies of West Pakistan, why should the same right of self-determined political status be denied to the Sikhs in India. A truncated Punjabi Suba, deprived of its natural resources was far from the expectations of the Sikhs.

The Resolution used a compromise formula to accommodate the demand of a 'self determined political status for the Sikhs in the united India' raised by Master Tara Singh group. It was also in consonance with the demand for more autonomy, put forward by several non-Hindi speaking states like Tamil Nadu, Kerala, West Bengal and others who had suffered on account of dominance of Hindi. This was also in keeping with the proclaimed aims, objectives and assurances by the founding fathers of the Constitution.

On December 11, 1972, the Working Committee of the Akali Dal formed a sub-committee of Sikh intellectuals to 'redraw the aims and objectives of the Sikh Panth and thus live up to the expectation of the Sikhs.' The chairman of the sub-committee was Surjit Singh Barnala. The moving figure behind this resolution was Sardar Kapoor Singh. The committee met on December 23, 1972, a draft was drawn up and was sent for approval to the working committee of the Akali Dal at Anandpur Sahib. The working committee gave its approval on October 16, 1973.

The Resolution demanded the restructuring of the Indian Constitution to ensure 'real federal principles with equal representation at the Centre for the states.' It suggested that the proposal of the Cabinet Mission, in 1946, be revived, with the Central Government in Delhi in charge only of Defence, Foreign Affairs, Currency and General Communications. The constitutional provision regarding the transfer of subjects from the state list to the concurrent list was considered an encroachment upon the rights of the states. Article 249 which empowered the Centre to legislate on subjects contained in the State List in the national interest was misused by the Centre to encroach upon the autonomy of the states.

Extra-constitutional institutions like the Planning Commission, Finance Commission, the Water and Power Commission and the University Grants Commission were further meant to curtail the powers of the states. Corporate bodies like the Cotton Corporation of India and Food Corporation of India also fortified the central hold over the states.

There was abundant erosion of federal content of the Constitution and unbalanced growth of unitary processes against the letter and spirit of the Constitution.

The Sikhs also resented the Central control over Punjab waters and energy through unconstitutional means. The idea of a joint capital and a joint High Court between the States of Punjab and Haryana, for an indefinite period, was also viewed as unprecedented and discriminatory. It was against this background that a demand for greater autonomy was reiterated in the Anandpur Sahib Resolution.

The issue of decentralization of powers, raised by the Anandpur Sahib Resolution, has engaged national attention. But the fact is that the demand for more state autonomy has been made by many other Indian States as well. So much so that the Tamil Nadu government formed a committee to suggest steps that, in the existing context, would make for a federal structure and ensure a full and rightful development of states. Its report, called the Rajmannar Report, was adopted by the Tamil Nadu Assembly, in 1971. This Report provides for a completely federal structure with autonomous states and with only four subjects viz, Defence, Foreign Affairs, Communications and Currency with the Centre. It advocated the classical form of federalism to make the Indian polity closer to that of the United States of America in the distribution of powers between the Centre and the States.

In a statement, made before the Sarkaria Commission, the Chief Minister of West Bengal also stressed the need for redefining Centre–State relations. He said, "Much of the tension and divisiveness currently afflicting the nation stem from factors which originate in imbalances in the existing relationship between the Centre and the federating states. Without a comprehensive restructuring of central–state relations, there is every fear that social and economic tension will multiply." He requested the Commission "to examine the provisions of such federal countries as the United States of America, Soviet Union, Canada and Australia and draw appropriate lessons" from their examples.

Jai Prakash Narayan's concept of 'Total Revolution' also envisaged more powers for the states, leaving only Defence, Communications, Currency and Foreign Relations with Centre. By and large, it reflected the same view point on centre–state relations, as contained in the Anandpur Sahib Resolution.

Rajamannar Committee Report

The Rajamannar Committee on Centre–state relations, set up by the Tamil Nadu government on 27 May 1971, recommended the constitution of a high-power commission to redistribute powers between the Centre and the states. The committee also suggested immediate formation of an ISC as provided for under Article 263 of the Constitution; abolition of the Planning Commission as at present constituted and its replacement by a wholly new one to be created with a statutory basis under a Parliamentary enactment and free from control by the central executive; widening of the states' share to include the corporation capital value of assets in the divisible pool; repeal of the Industries (Development and Regulation) Act of 1951, with a new Act providing for control by the Centre of only industries of national or all-India character; and conferment of industrial licensing powers on the states.

Sarkaria Commission

A commission to look into Centre–state relations was appointed on 9 June 1983, by the late Prime Minister, Mrs Indira Gandhi, in the wake of the Telugu Desam movement which reduced the position of the Congress in the state to shambles. It was headed by Justice R.S. Sarkaria.

The limits that the government wanted to impose on the commission's investigation were strongly criticised by the Opposition. According to the terms laid down, in making of its recommendation, the Commission:

> ...will keep in view the social and economic developments that have taken place over the years and have due regard to the scheme and framework of the Constitution which the founding fathers have so sedulously designed to protect the independence and ensure the unity and integrity of the country which is of paramount importance for promoting the welfare of the people.[6]

Justice R.S. Sarkaria presented the report of the Commission—headed by him—on Centre–state relations, to Prime Minister Rajiv Gandhi, in New Delhi on 27 October 1987. The sizeable document—a product of four years of labour—contains recommendations of changes in the existing arrangements between the Centre and the states necessary, in its view, to remedy the problems arising from the present relationship.

In its unanimous report, the Sarkaria Commission said that there was no need to amend the Constitution to give more powers to the states because it already had provisions to allow them freedom in their spheres.

The Commission, however, pointed out that the Centre had been usurping the states' prerogatives, poaching on their sphere and violating the letter and spirit of the Constitution by expanding the Concurrent List at the expense of the State List of subjects.

Chelliah

In 1991, Dr Raja Chelliah of the Fiscal Research Foundation, New Delhi, delivered the fourth L.K. Jha Memorial Lecture. The subject of the lecture was 'Towards a Decentralised Polity'. Chelliah was the main architect of tax reforms when economic reforms were initiated in 1991. Chelliah's approach focused on more equitable revenue-sharing between the Centre and the states. His remarks were adopted as policy platforms by several regional parties of south India.

---

[6]For more details, visit the website of the ISC at http://interstatecouncil.nic.in/SARCOMM. htm

Karunanidhi

Dr M. Karunanidhi, at that time the Chief Minister of Tamil Nadu and the president of the Dravida Munnetra Kazhagam, and a man who had substantially shaped the politics of his state, argued for state autonomy and federalism at the Centre. In an article in the *The Hindu* (dated 15 August 2007), he stated:

It needs to be remembered that only the spirit of 'co-operative federalism'—and not an attitude of dominance or superiority—can preserve the balance between the Union and the States and promote the good of the people. Under our constitutional system, no single entity can claim superiority. Sovereignty does not lie in any one institution or in any one wing of the government. The power of governance is distributed in several organs and institutions—a sine qua non for good governance. Even if we assume that the Centre has been given a certain dominance over the States, that dominance should be used strictly for the purpose intended, not for oblique purposes. An unusual and extraordinary power like the one contained in Article 356 cannot be employed for furthering the prospects of a political party or to destabilise a duly elected government and a duly constituted Legislative Assembly. The consequences of such improper use may not be evident immediately. But those do not go without any effect. Their consequences become evident in the long run and may be irreversible.

The determination of the extent required and feasible decentralisation is crucial. At one extreme (counting out the secessionists) the demand has been put forward that the only central responsibilities should be defence, foreign affairs, currency, communications. This was not only the demand of the Anandpur Sahib resolution of the Akali Dal but also that of the Asom Gana Parishad, the Anna Dravida Munnetra Kazhagam and several other parties. When Biju Patnaik was Chief Minister of Orissa he too enunciated a similar position. His emphasis was on 'fiscal independence' for the states, including the right to conduct foreign economic relations such as obtaining loans and investments from abroad.

A synopsis of the positions of the Rajamannar Commission, the views of Dr Raja Chelliah, the position of the Anna Dravida Munnetra Kazhagam and the Justice R.S. Sarkaria Commission are given in the Appendix along with an analysis by Prof. Dilip Singh of the Department of Political Science, Punjabi University Patiala.

7

Punjab versus the Central Government Steamroller

INTRODUCTION

In the case of Punjab, the issue of autonomy has a religious angle. The Sikhs constitute only 2 per cent of India's total population but the majority of Punjab's population is Sikh. The Sikhs are keenly aware of the difficulty of maintaining their unique identity. The Hindu ethos strongly pervades everything from school textbooks to popular films and television series and advertisements. They are not against 'national integration' but they see no reason why they should have to assimilate with the national majority and become indistinguishable from it. What gives added sharpness to the feeling that the majority is over-demanding is the fact that during the freedom struggle and in subsequent wars, it has been the Sikhs who made maximum sacrifices for the country.

For many years in the post-Independence period, the Congress Party dominated the Centre and the states and this meant that issues relating to power-sharing did not arise and were not discussed. However, in the past two decades, as different parties formed governments in the states and the Centre and several regional parties have come to power, the issue relating to power for the states has assumed significance.

Attempts to raise issues pertaining to state powers have been treated as somehow, anti-national per se, although more and more states would like a debate on this. Punjab is not alone in complaining of central apathy and demanding more powers to decide issues which are of immediate concern. Many other states, want more powers for the sake of faster development, leaving only matters relating to foreign affairs, defence, currency and communication to the Centre.

Over the years, the Centre has gradually assumed more powers and intruded into areas that are essentially states concerns, if one goes strictly by the division of powers laid down by the Constitution.

As laid down by the architects of India's Constitution, the states play a determining role in development of agriculture, water resources, land relations, environment and forests, rural roads and state highways, minor ports, electricity, and rural and urban services. They are also responsible for human development through investments in key areas such as health and education. To enable them to fulfil their responsibilities, the Constitution also empowers them to raise taxes, get a share in funds available with the Centre, and levy user charges on various economic and social services provided to the people. The Constitution also provides for rights and responsibilities of local bodies like Panchayats and municipal bodies. The Constitution envisages that the states are as much, if not more, responsible for development as the Centre.

As explained in the previous chapter, shortly after Independence, the Centre began to encroach on the powers and autonomy of the states. The result was that the Centre sought to assume a larger responsibility for development than assigned to it in the Constitution. It floated centrally sponsored schemes and procrastinated on sharing all the tax revenues with the state governments. The states were required to seek clearances from the Centre for setting up projects for infrastructure and human development.

The Centre has starved Punjab for years. The *Economic Survey of India 1980–81* reported:

> For every rupee spent on Plan programmes in Punjab, the central government contributes only 15 paise. Compared to this meagre amount, Kerala receives 43 paise, Bihar 42 paise, Andhra Pradesh 20 paise, Tamil Nadu 30 paise, Uttar Pradesh 31 paise and West Bengal 20 paise.[1]

---

[1]For more details on state assistance under the Five Year Plans, see *The Economic Survey of India 1980–81.*

Discrimination against Punjab on the economic front was brought out by I.K. Gujral, former Prime Minister and Foreign and Planning Minister. He stated:

> [...]Punjab has been a victim of two myths: one, as Indira Gandhi put it, 'It is an advanced and prosperous state' [...] And two, 'that the people of Punjab are well above the poverty line since its per capita income is the highest in the country'.[2]

LOW INVESTMENT IN PUBLIC SECTOR

Gujral further asserted that Punjab has had an unfair deal in the setting up of industries. He states that, 'During my brief tenure as a Minister in Planning Ministry in 1976, five additional mills [sugar] had been sanctioned. Subsequently after my departure for Moscow, these were re-allocated to some adjoining states.'[3] In sum, only one mill was sanctioned in the decade 1970–80 when the sugarcane production was annually increasing at the rate of 2.54 per cent compound. As Figure 7.1 shows, only about 2 per cent of India's agricultural produce is processed.

Public sector investments too have been at the lowest in Punjab resulting in industrial stagnation. The Industry Minister, Ajit Singh, stated in the Lok Sabha:

> Maharashtra had the highest central public sector investment and Punjab the lowest in the past ten years. While investments in Maharashtra were to the tune of ₹15,150 crore, in Punjab they were ₹457 crore. The total investment in the public sector projects, numbering 70, was ₹80,912 crore, in the past ten years.[4]

Thus, the share of the total investment in the public sector projects in Punjab comes to less than 0.4 per cent as against 18.72 per cent for Maharashtra.

---

[2] I.K. Gujral, 'The Economic Dimension', in *Punjab in Indian Politics: Issues and Trends*, ed. Amrik Singh (London : Oriental University Press, 1986), 44.
[3] Ibid.
[4] Lok Sabha Debates, 11 April 1990.

Figure 7.1: **Country-wise comparison of value addition through food-processing**

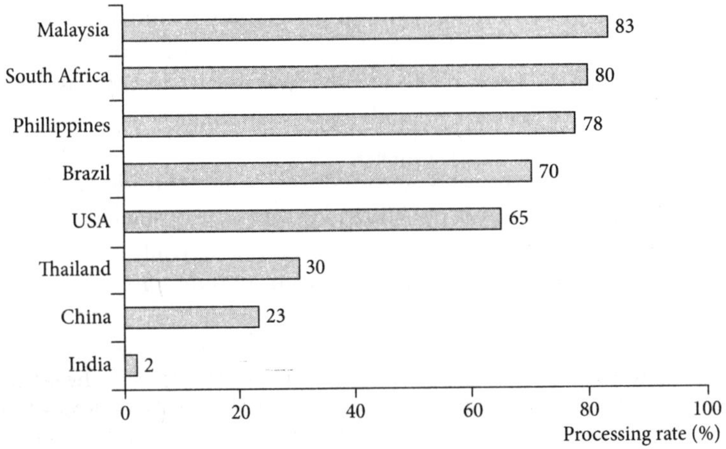

DENIAL OF INDUSTRIAL LICENSING

Punjab's share of licenses has almost been negligible. In fact, there was a back-slide in the industrial position of Punjab between 1965–66 and 1977–78. In 1965–66, Punjab ranked eighth with 4.1 per cent of the share of value added; in 1977–78, it had come down to the tenth position accounting for only 2.8 per cent of the share of all-India value added. The position has worsened further since 1977–78. I.K. Gujral points out: 'The Punjabi protest is against being treated as producers of raw materials while value added benefits go to others.'[5]

Punjab grows 22 per cent of the total cotton produced in the country. But, only 3.3 per cent is being processed in the state and the rest goes out raw to other states for processing. In order to process just 50 per cent of total production in the state, there is a need to have at least 27 mills, each with a 25,000 spindle capacity. Punjab sugarcane growers are able to sell only 13 per cent of their produce to the mills. The all-India average is 33 per cent. It is

[5]I.K. Gujral, 'The Economic Dimension', in *Punjab in Indian Politics: Issues and Trends*, ed. Amrik Singh (London : Oriental University Press, 1986), 45.

90 per cent in Gujarat, 82 per cent in Maharashtra, 50 per cent in Bihar, and so on.

In order to convert Punjab into the granary of India, the Centre denied the state heavy industry. The state's vulnerability on account of its location on the border with Pakistan was used to justify this denial. This might have been true decades ago when military technology was primitive, but as both countries acquired air and nuclear capability and missiles, it ceased to be valid. Agro-industry is often considered ideal for Punjab but even this was not allowed except as very small-scale units. Punjab has missed out on value addition for its agricultural produce.

According to Shekhar Gupta, the editor of *Indian Express*:

> Mumbai runs on business and enterprise, Punjab survives on agriculture which is no longer quite the same driving force as modern enterprise. Punjab's is the story of a state that rode the crest of the green revolution but missed the industrial revolution altogether. The greatest tragedy of the post-insurgency years is that nobody has addressed that issue and the result is stagnation, joblessness, and bankruptcy.[6]

Concerned over the ever-decreasing tax collections, the Punjab government filed a suit in the Supreme Court in 2009 challenging the central government's tax exemption given to industries in the neighbouring states of Himachal Pradesh, Jammu and Kashmir and Uttarakhand. The discriminatory fiscal incentives in form of excise duty, income tax benefits and investment subsidies to industries in the three states had led to exodus of industrial units from Punjab.

In spite of the desperation in the agro-sector, the central government appears to be determined to retain Punjab as the bread basket of India and deny alternate sources of development and employment.

STRANGLEHOLD ON AGRICULTURE

The farming community, which is increasingly asserting its interests, finds powers of administration at the state level being continuously

---

[6]Shekhar Gupta, 'From Guns to Roses, and Now a State in the Red: The Punjab Parable', *Indian Express*, 19 August 2000.

eroded by the central government. The growing economic and social role of the administration is being monopolised by the central government. Whether it is economic planning and diversification of financial institutions or the radio-television communication network, it is all under the monopoly of the central administration. In this situation, promoting its class interest through administration at the state level has become extremely limited.

MANIPULATION OF PRICES

Another agency, which advises the central government on prices of industrial goods is the Bureau of Industrial Costs and Prices. The central government intervenes in the pricing of various commodities and has the capacity to change relative prices. Over the last several years, the price system has weighted against agriculture and was in favour of industry under the pressure of Indian monopoly bourgeoisie which has a decisive influence over the central government. The Terms of Trade (TOT) have moved against agriculture from 100 in 1970–71 to 81.8 in 1980–81.[7] The fall is sharp and consistent from 1974–75. A major part of the surplus wheat, paddy and cotton is procured by central agencies such as Food Corporation of India (FCI) and Cotton Corporation of India (CCI), and inputs such as fertilisers, pesticides, insecticides, weedicides, tractors, pumping sets and electric motors are supplied by Indian—private and government owned—and foreign monopolies.

The massive involvement of the Centre in the affairs of the state is made possible by an enormous amount of financial resources at its disposal. Of the total budgetary resources of the central and state governments, more than two-thirds first accrue to the Centre. Expenditure is made 50–50 by both the Centre and the states. Thus, states depend on the Centre for their resources to the extent of one-third. This makes the position of the state

---

[7]Sukhpal Singh in *Economic Development and Structural Change in Punjab: Some Policy Issues*, ed. Sucha Singh Gill (Ludhiana: Principal Iqbal Singh Memorial Trust, 1994).

administration very precarious, particularly when a state is ruled by an opposition party.

The states have little or no independent powers for contracting loans. Since the Centre has an exclusive power to obtain foreign aid or local loans, or draw on deposits with the nationalised banking and financial institutions, the states remain completely at the mercy of the Centre in the allocation of funds.

CREDIT-DEPOSIT RATIO

An important measure of economic exploitation of Punjab is the credit-deposit ratio which is the lowest in the country. The resources mobilised by the commercial banks from Punjab are not utilised in the state. Punjab's banks receive huge remittances from over one million Sikhs, who are either permanently settled abroad or go abroad temporarily to earn. Bank deposits in Punjab are quite high but ironically 70 per cent of the deposits are diverted for development to areas outside the state. In other words, the state has capital but is not allowed to use it for its own people.

TRIPLE EXPLOITATION

In relation to Punjab, the Indian government has pursued the triple policy of exploitation. First, it is heavily draining out its water and hydel power to erode the base of all agricultural and industrial development. Second, its financial resources are being largely diverted to other states. Third, it is seen that pro- cessing industries for the raw materials produced in the state, and for which it has a natural advantage, are not developed in the state. Punjab like any colony, continues to supply only raw materials to the faraway industries of the country. The economy of Punjab has been shattered.

CORE ISSUES

River Waters

Punjab is likely to become a desert in the next 30 years as the ground water level in the state is falling at a rate of 23 centimetres per year for the last 15 years.

Punjab's Water Demand

In 1990, Punjab Agricultural University (PAU) experts worked out the normative water demand for the crops grown in Punjab as 37.6 MAF, and adding this to the quantity for civic needs, the demand rises well above 43.25 MAF.

Gap between Availability and Demand

The gap in water availability works out to 22.75 MAF and this gap is being met by tapping ground water—which means 'mining' or 'over-exploiting' of the ground water resource—resulting in problems in different areas. PAU has warned that it took only 20 years (from 1982 to 2001) for the water table to deplete by an average of 4 metres. Further, indiscriminate water usage has seen the water table deplete from 25 centimetres to 3.53 metres in the last 4 years alone. 'The problem we are facing in the farm sector', it says:

> ... is not only of the depleting water table, but also of bad quality of water being used for irrigation ... Yet another equally worrisome possibility could be that when a sizeable quantity of presently good quality water gets mined out from the upper ground water strata, the subsequent lower level supplies may turn out to be unfit for irrigation.[8]

Punjab is heading for self-destruction because of the excessive use of ground water for agricultural and other purposes. This is indicated in the mapping done by the local regional office of the

---

[8]S.S. Prihar, S.D. Khepar, Raghbir Singh, S.S. Grewal, S.K. Sodhi, *Water Resources of Punjab* (Ludhiana: Punjab Agricultural University, 1990).

Central Ground Water Board with regard to the use of ground water in the state. Of the 137 development blocks mapped by the board's scientists, 103 are over-exploited as far as the use of subsoil water is concerned. Five blocks are at a critical stage and four at a semi-critical stage. Blocks falling in the safe zone number 25. Ironically the safe zone consists mostly of those blocks where the subsoil water is unfit for drinking as well as for irrigation. Simply put, it means that if 100 litres of water goes into the subsoil as replenishment in these particular blocks, 200 litres are extracted through tube-wells and other means. There are about 1.2 million tube-wells, mostly operated on electricity, in Punjab.

A recent study has produced evidence that large tracts of land in the state are being irrigated with poor quality water. It forewarns against the continuous use of such water as it will lead to a drop in the crop yield—the biggest source of income and the backbone of the economy in this 'granary of India'.

After it had been decided to give 8.0 M.A.F. of Ravi–Beas waters to Rajasthan, P.S. Kumedan—Punjab's river water case expert—pointed out at a conference held on 29 January 1955—under the chairmanship of Gulzari Lal Nanda, Union Irrigation and Power Minister—that this water was not to be given free of cost. As the conference was concerned only with the distribution of water, it was decided that the cost of water would be worked out separately by the states. The cost of water going to Rajasthan has not been calculated so far. About 400 MAF of water has gone to Rajasthan from Punjab's river during the last 40 years. The cost should be worked on the escalating cost of water. Besides this, Punjab has to be compensated for the depletion of 400 MAF acre feet of its sub-soil water on this account and also the cost of electricity for pumping out this water.

Kumedan pointed out:

> Last year, Central Water and Power Commission issued a pressnote stating that the iron gates of Madhopur Headworks had become very old with the result that 100 cusec water was leaking through these gates every day and it was going waste to Pakistan. They put the annual value of this 100 cusecs at ₹100 crores, which loss Punjab was suffering every year. Punjab is giving one crore acre feet or say 50 lakh cusecs of water to Rajasthan every year (1 cusec day = 2 acre feet). On this basis, the value of this much water comes to about ₹14,000 crores annually.

The total value of 40 crore acre feet of water supplied to Rajasthan during the last 40 years would thus come to ₹5,60,000 crore.[9]

The biggest loss suffered by Punjab for supplying 10 MAF canal water to Rajasthan is that it has to extract this much extra groundwater for its own use. There are more than 13 lakh electric and diesel-operated tube-wells in Punjab which pump out about 2.50 crore acre feet of ground water every year. As per Punjab State Electricity Regulatory Commission (PSERC), electricity consumed by these tube-wells is more than 1,000 crore units every year and the value of this electricity, at the rate of ₹2.50 per unit comes to about ₹2,600 crore. However, as Punjab purchases electricity from other states at the rate of ₹7 or 8 per unit, at this rate, cost of this much electricity would come to more than ₹7,000 crores. Diesel-operated tube well is four to five times more costly than an electric one. Due to shortage of electricity, many farmers use generators as well. Taking all these factors into consideration, if we assume ₹5 per unit as the cost of electricity, the total cost of 1,000 crore units of electricity would come to ₹5,000 crore.

Diversion of Punjab river waters to the non-riparian state of Rajasthan is not made in the national interest. Taking river water through open canals to distant Rajasthan results in considerable evaporation loss on the way and a large quantity of water is wasted through seepage in the sandy soils of the desert. Even the management of the Bhakra Management Board has been taken away from Punjab.

In 1988, flood water had to be released from the Bhakra reservoir at a time when the downstream area was already flooded. Official estimates put the flood damage in 1988 at ₹2,700 crores in addition to a loss of 700 human lives. Pakistan reported sightings of 1,700 bodies downstream. Unofficial estimates placed the damage at ₹50,000 crore and anything over 5,000 human lives. For this huge loss, the central government and the beneficiary states paid zero compensation to the people of Punjab.

There are 800 weak points in the total 225 kilometres stretch of the Bhakra Main Line (BML) canal in Punjab. Parvesh Sharma of the *Times News Network* reports that 'there has been no repair

---

[9]Letter by P.S. Kumedan, (PCS, Irrigation Advisor to the Punjab Government) to Captain Amarinder Singh (Chief Minister of Punjab) dated 28 December 2008.

ever since 1992 when technically it should happen every two years'.[10]

The failure of the state authorities in repairing the canals may wreak havoc in the state, as at many points, cracks have surfaced and the cement has fallen off. Sources in the BML said 150 weak points alone were reported to be in Fatehgarh Sahib district. For this reason only 7,500 cusecs of water had been released despite a capacity of releasing 12,500 cusecs.

D.S. Sandhu, superintendent engineer of the BML, while admitting that this could lead to a disaster in the state said, 'In August end we have a meeting with other states and hope that they would understand our problem'.[11] With no one ready to own up responsibility, the canal has turned into a picture of neglect. Summing up the problem, a senior BML official said, 'Other states are unwilling to contribute money as per their share of water but in case of a mishap Punjab would suffer the most'.[12]

Punjab gets only 10 per cent of its water from the BML for irrigation purposes while Haryana gets 55 per cent and Rajasthan 30 per cent; the remaining water goes to Delhi for drinking purposes. All states are bound to contribute for the repairs but they avoid payment.

The central government now appears to be changing its policy on interlinking of rivers. 'Less than a month after Rahul Gandhi warned against "playing with nature", Union Minister for Environment and Forest, Jairam Ramesh, said the idea of interlinking India's rivers was a "disaster"' (*Indian Express*, 5 October 2009).

Taking river water beyond the river basin area has similar consequences as the linking of rivers. Punjab has suffered colossal damage due to transfer of Punjab river water to non-riparian and non-basin states. Punjab deserves to receive reparation for the loss suffered by the state. Future damage should be contained through gradual withdrawal of river water supplied to the other states.

Land Ceiling

Dr S.S. Johl, an eminent agro-economist, says that the Land Ceiling Act in India spelt death for the agriculture sector. The purpose for

---

[10]Parvesh Sharma, '800 Weak Points in Bhakra Main Line', *Times News Network*, 16 August 2008.
[11]Ibid.
[12]Ibid.

which it was created has lost significance. Gone are the days of the *zamindari* system. There is now no limit on the size of industry and scale of business that can be operated; it is only the agricultural holdings that are subject to ceilings. As per the Land Ceiling Act of 1972, in Punjab the ceiling for farming family is only 7 hectares of double-cropped irrigated land and 11 hectares for single-cropped irrigated land. For unirrigated land, the ceiling is at 22 hectares (55 acres). In fact the agriculture sector has been used as a sort of guinea pig for all the misplaced socialistic dispensations in the country. The redundant farm labour has been kept bottled-up in the agriculture sector through low ceilings and distribution of small uneconomic parcels of so-called surplus land. It has not been realised that farming is also a business that involves costs-returns considerations, factor use efficiency and market competition. Productivity per unit of land operated is only one of the considerations that is important from the point of view of total production. Small farms have high productivity in terms of the land utilized. Although new production technology is now providing an edge to larger farms—even if this assumption of small farms having higher productivity is assumed to be true—there is an important consideration of viability of farm size. Farms must be big enough to survive.

The farming business can remain viable only if farm units grow through internal capital formation and adoption of improved fast-changing production technology to become cost-effective and competitive in the national, as well as international market. Unviable small and marginal holdings might have had a place in a socialistic production system that seldom operates on economic principles but not in a free competitive market system. The way the economy is being globalised, such small uneconomic holdings have no place. Today even a 7 hectare two-crop irrigated farm yields a net income for the farm family—including imputed wages of family labour—that is less than the annual income of the lowest paid government employee. Moreover, the government employee's livelihood is free from any risk. Besides, educational and health costs of comparable service are much higher for rural families. The agricultural entrepreneur in the farm sector has to be, therefore, freed from the shackles of stunting state controls on the size and scale of agricultural enterprise, in order to let the farm enterprise grow strong enough to face market competition at the national and international levels.

Unless agriculture as a free business/industry attracts capital investments as well as educated and informed entrepreneurs, and becomes capable of using modern technology, it cannot start traversing any perceptible growth path that may generate agricultural surpluses at low cost to compete in the market. This will create additional avenues for gainful employment and generate wage goods that will match the outflow of labour from the farm sector. It is, therefore, essential that the Land Ceilings Act be reviewed to provide the much needed impetus to growth and development in the agriculture sector.

Under the accelerated process of economic growth, there must occur transfer of population from the farm sector to the non-farm sectors. Unfortunately, in India and also in Punjab, sufficient shift of population to the non-farming professions has not taken place. Instead, absolute number of people dependent upon agriculture has increased in the state as in the country at large. Between 1980–81 and 1990–91, about one lakh new operational holdings have come into being in Punjab, and almost all of these holdings are marginal holdings. On the social front, inequalities in the rural sector may get accentuated initially, but very soon the soothing effect of the policy will begin to show with the expansion of better paying on-farm and off-farm employment opportunities.

Earlier, in the era of traditional technology, the farm productivity was higher on small holdings, but now there is not much difference in productivity between small and large holdings.

To ensure that the rural sector was not unduly perturbed by imposing a land ceiling on farms, urban land ceiling was also proposed under the Ceiling and Regulating Act, 1976. Initially, states of Andhra Pradesh, Haryana, Himachal Pradesh, Karnataka, Maharashtra, Orissa, Punjab, Tripura, Uttar Pradesh and West Bengal adopted the Act. Under this Act, ceiling was imposed on land clubbed under various categories of towns. In the A class categories, i.e. the cities, the ceiling on vacant land was placed at 500 square metres. In the other categories, it was proportionally more. However, as it transpired later, the urban land ceiling drama was created merely to soften the farming sector. This urban Land Ceiling Act was subsequently repealed. The Repeal Act was enforced in the state of Haryana, Punjab, Uttar Pradesh, Gujarat, Karnataka, Madhya Pradesh, Rajasthan, Orissa and all the union territories.

UNION AGRICULTURE POLICY

From Jai Kisan! to Hai Kisan!

The direction of the Centre's policy on farming is to take agriculture out of the hands of farmers and place it in the hands of large corporations. Union budgets and legislations consistently push this idea further forward. We are witnessing the largest displacement in our history. It is not happening in a dam or a mining project. It is happening in agriculture. Central policy is making farming impossible for small holders. Millions are being forced out of agriculture and their traditional way of life and they have no place to go.

P. Sainath has aptly pointed out:

> [...] sixty years on, rural India is a shambles. The most severe agrarian crisis since the eve of the Green Revolution rages on, but does not hold elite or media interest for long. Farm incomes have collapsed. Hunger has grown very fast. Public investment in agriculture shrank to nothing a long time ago. Employment has collapsed. Non-farm employment has stagnated.[13]

There is an urgent need to hike land ceiling substantially with a provision that only bonafide agriculturists of the state can buy agricultural property. The neighbouring states of Himachal Pradesh and Rajasthan have added this safeguard.

Costs and Prices

On cost and prices, the article by Harish Damodaran which clearly brings out the falling profitability of farming, the drying up of non-farming opportunities and the growing fragmentation of landholdings all make agriculture is a good reference point.[14]

For data on average yields and costs, one may refer to the Commission for Agricultural Costs and Prices' (CACP) latest estimates for the 2005–06 sowing season. These numbers, particularly

---

[13]P. Sainath, 'The Decade of Our Discontent', *The Hindu*, 9 August 2007.
[14]The said article may be found at Harish Damodaran, 'Why Farming Has Become Unviable', *Hindu Business Line*, 27 July 2006.

relating to costs, tend to be on the lower side and to that extent, exaggerate the actual returns accruing to farmers. However, they give an approximate picture.

A report published by the Department of Economics and Sociology, Punjab Agricultural University (April, 2009), points out that

> during 2001–05, the minimum support prices of wheat and paddy were almost frozen due to falling international prices. The minimum support price of wheat increased by about 1.5 per cent per annum and that of paddy by about 2.5 per cent during this period whereas the cost of production of wheat and paddy went up by about 8–9 per cent per annum. Further, the productivity of wheat declined during 2001–05.[15]

As a consequence, the economic distress of farmers in the state increased manifold leading to the phenomenon of farmers' suicides. What can be said about the repaying of debts when their incomes were not even sufficient for meeting subsistence expenditure.

Even after factoring in the net income from milk and sale of by-products such as straw (say, ₹1,000), the maximum that the farmers in the country's well-endowed areas can earn would be around ₹3,000 a month. This is when one is assuming that there are no crop losses due to hailstorms, floods or pest attacks.

Three-fourths of Indian farmers take home less than ₹3,000 a month, a figure that is roughly 60 per cent of the starting salary for a government office attendant! According to National Sample Survey (NSS) Data (2003) the average total income of farm households with up to 2 hectares was less than 80 per cent of their consumption expenditure.

The Chief Minister of Punjab, Parkash Singh Badal, lamented in the State Assembly that the annual earning of a farmer is around ₹12,950, which means that his daily earning is ₹36; much below than what menial workers of other classes get.[16]

The Planning Commission has projected Punjab as one of the four slowest-growing states in the country with less than 4 per cent growth recorded, as against the average of 6 per cent at the national level.

---

[15] *Market Intelligence Bulletin*, Department of Economics and Sociology, Punjab Agricultural University, Ludhiana, April 2009.

[16] Parkash Singh Badal quoted in 'Even Our Peons Earn Better Than Farmers', *Times of India* (Chandigarh edition) 20 March 1999.

The state government has argued before the Planning Commission that at ₹35,000 crore, Punjab had the highest level of farm indebtedness in the country. In terms of average per household, it comes out to be ₹41,576, which is far greater than the national average of ₹12,585. Agro-economist M.S. Swaminathan has suggested that the Minimum Support Prices (MSP) should be 50 per cent plus the costs incurred by farmers, or it should be linked to the Consumer Price Index.

Time News Network on 2 August 2009, reported that a state-level committee headed by economists Ranjit Singh Ghuman (Chairman), Punjabi University; Professor P.S. Rangi, Consultant to the Farmers Commission; and Dr M.S. Sidhu, Head of the Economics Department, Punjab Agricultural University, in its report to the Punjab government on un-remunerative MSP, revealed the crop-wise break-up of loss: ₹41,411 crore on account of wheat alone, with an annual average loss per farmer of ₹1,035, and the paddy farmers lost ₹202.85 crore, with an average loss per farmer of ₹507 crore annually, due to non-linking of the MSP of wheat and paddy with the Wholesale Price Index (WPI). The report supplied data on Punjab's year-wise contribution to the Central Pool (see Table 7.1).

But even as India became a food-sufficient country, thanks to Punjab's contribution, the state suffered a huge loss in terms of serious erosion of its soil fertility, depletion of sub-soil water and deteriorating ecological balance.

Table 7.1: Punjab's share in central pool

Wheat		Rice	
Year	*Contribution (%)*	*Year*	*Contribution (%)*
1968–69	79	1968–69	7.36
1968–70	74	1970–71	16.42
1970s	53–63	1977–78	45.13
1980s	54–72	1978–91	37–60
1990s	51–71	1990s	33–43
2001–07	52–75	2000 onwards	29–38

Source: *Losses Suffered by Farmers of Punjab since 1967–68 Due to Non-linking of the MSP of Wheat and Paddy with WPI*, Report submitted to the Government of Punjab by Ranjit Singh Ghuman (Chairman), P.S. Rangi (Consultant to Farmer Commission) and M.S. Sidhu (Head of the Economics Department), Punjab Agricultural University, Punjab.

Unfavourable Price Structure

Punjab has suffered decades of unrealistically low government-set MSP and high input costs. Consumers of other states have enjoyed subsidised prices at the cost of Punjab's farmers for decades. Punjab and Haryana combined contribute approximately 67 per cent of all wheat procured in India. Overall contribution (all grains) comes to around 60 per cent. The largest producers are automatically the largest sufferers.

It would be better to relate MSP to the market price index till such time that farmers are in a position to match import price. Likewise, a buffer is needed against influx of cheap imports under the new World Trade Organization regime.

Another important measure would be to fix MSP for crops selected for diversification.

Rural Debt

According to a recently conducted study by Professor H.S. Shergill,[17] Director of Research, Punjab Development Studies at the Institute for Development and Communication, Chandigarh farm debt in the state has increased five times over in the last decade. The per farm household debt has tripled in the 10 years, from ₹52,000 per household in 1997 to ₹1.39 lakh in 2008. Further, per acre amount of debt has more than doubled over the same period from ₹5,721 to ₹13,062.

Nearly 72 per cent of farm households are heavily involved in debt. Out of these around 17 per cent are in a virtual debt trap, in the sense that they cannot pay even the annual interest on their loans from their current farm income. Shergill has said there is little chance of their repaying the accumulated debt from the current income.

Shergill, a well-known economist and a former colleague of Prime Minister Dr Manmohan Singh, has concluded that the outstanding debt component has increased at a faster rate (14.13 per cent per year) than total farm debt (8.81 per cent per year) over this period. The mortgage debt, however, has declined over this period and may completely disappear in the near future. Interestingly, the debt

---

[17]Article in *The Tribune* (2 December 2009).

of small and marginal farmers has grown at a slower rate (1.29 per cent per year) than the debt of medium and big farmers (2.71 per cent per year).

Almost 60 per cent of these debt-trapped farm households are marginal and small farmers and most of these (86 per cent) belong to the Malwa region. When compared to income generated from the farms, the debt amount has increased from being 68 per cent in 1997 to 84 per cent in 2008. Then as a proportion of the value of machinery owned by Punjab farmers, the debt amount has gone up from being 15 per cent in 1997 to 53 per cent in 2008. Despite a steep rise in farmland prices in the state, the amount of the farm debt as per data in 2008 is equal to 4 per cent of the total value of farmland of the state, as compared to the 3 per cent figure in 1997.

Debt Trap

- Average debt per farm household is ₹1,39,000.
- Seventy-two per cent of farm households are heavily involved in debt.
- Seventeen per cent cannot pay back even the interest.
- Sixty per cent of the debt–trapped farmers are small or marginal farmers.

Since 1988, Movement Against State Repression (MASR) has been documenting cases of suicide in Moonak and Lehra subdivisions. MASR research covering data from the 1990s and the first decade of the 21st century two decades, has found 1,630 suicide cases in 91 village of Moonak and Lehra subdivisions. Of this total, the panchayats have given 1,570 affidavits verifying the facts of these cases.

In every single case, the victim or the father or the eldest brother or the husband, was deeply in debt. In most cases, the victim's landholding was very small. In some cases, the victim committed suicide immediately following the loss of agricultural land. Sometimes land loss was gradual. The land was sold off in small parcels, year after year, and when the last small bit was lost, the farmer now bereft of livelihood, took his own life. Sometimes the land had been hypothecated to the moneylender and when the debtor failed to repay his loan, the moneylender foreclosed and took the land.

On at least three occasions, the former Chief Minister Captain Amarinder Singh publicly promised to introduce the non-institutional debt relief Bill that stood vetted by the Revenue Department. He set up a high-powered committee comprising Rajinder Kaur Bhatal, Lal Singh and Surinder Singla—three senior-most ministers—but the powerful *arthiya* lobby managed to scuttle it each time.

It is the same under the present Shiromoni Akali Dal (SAD)–Bharatiya Janata Party (BJP) government. In the years that it has been in power in the state, the government has held over half-a-dozen meetings to discuss the draft Bill, but with no concrete results.

Paddy and Wheat: The Water Table Problem

The Punjab government must realise that in producing food for the nation, it is losing dearly in terms of its non-renewable natural resources for which there can hardly be any compensation.

Sudhirendar Sharma, a well-known development journalist, points out that Punjab literally feeds India with its annual contribution of 53 per cent of wheat and 40 per cent of paddy to the country's food stocks. He contrasts this with the position in 1947 when India was a food grain deficit area with only 52 per cent of its area under irrigation.

The 1960s Green Revolution changed all of that. Introduction of dwarf wheat germ-plasm and dwarf varieties for paddy crop resulted in quantum leaps in production. The high tide of the Green Revolution led to intensive production which then led to the emergence of crop monocultures, in general, as farmers, enticed by the productivity of the High Yielding Variety (HYV) seeds promoted by the agricultural establishments of the country, switched to rice and wheat rotation. Pulses and coarse grains were sidelined; paddy and wheat was where the money was and to these crops over 71 per cent of the gross cultivated area was put. Farm machinery, pesticides, fertilisers and irrigation dramatically increased the productivity of land. Today 95 per cent of the net sown area gets irrigated by a web of canals and tube-wells.

But land as a factor of production has its limitations and is cannot support the intensity of such agricultural practices for long. Crop yields and water resources have started to decline steeply as a result of it. Realising that the ecological threat was real and close to home, the state government now talks of weaning away farmers from such paddy–wheat cropping patterns. With stagnant growth rates of 73 per cent, Punjab has been forced to undertake this shift in farming practices. It is now seeking a support of ₹1,280 crores from the central government. It needs this money to compensate its farmers for switching from the traditional cropping system. By giving an incentive of ₹12,500 per hectare, the state hopes to be able to relieve some 1 million hectares under paddy–wheat rotation to be replaced by alternate crops like pulses and oilseeds.

For the central government, it is the bigger question of feeding the country, should the flow of wheat and paddy stop flowing from this machine. However, for Punjab it is a question of ecological survival—of sustaining its natural resources like water and soil in a healthy state. Politically and economically it is also a question of the sheer survival of its farming community. Change in crop patterns for Punjab will save the country an estimated ₹8,976 crore in procurement, handling and storage costs. It is also intended to provide some relief from the high incidence of pests, diseases and hosts of ecological problems, like the alarming depletion in the water table, water logging, soil salinity, toxicity and micronutrient deficiency. These are the less-talked about afflictions of the HYV boom, which was once seen as the engine of food security and the driving force behind farm prosperity.

A reduction in paddy and wheat production by 30 per cent is being suggested as the antidote to the current stress on the state's water resources. Rice is not a traditional crop in a semi-arid state like Punjab. Rice fields alone consume some 85 per cent of all fresh-water supply. This adds up to a total annual consumption of 44 lakh hectare metres of water. Being a water-guzzler, paddy is the key crop being targeted. Undoubtedly, a shift in the cropping pattern will ease the pressure on the already over-stretched groundwater resources in the state.

Not surprisingly, water tables have long dropped beyond the reach of muscle-driven water lifts, dipping down to as low as 400 to 450 feet in many places. As a result, some 84 development blocks—out of the 138 in the state—have already been declared as being in the dark zone where the level of groundwater exploitation is over 98 per cent, as against the critical level of 80 per cent. Six out of the 12 districts in the state have recorded a groundwater utilisation rate of over 100 per cent. In many parts of Punjab, the water table is falling at the rate of 1 metre per year.

This crisis of overuse is the ostensible reason for the Punjab government's measure to reverse the existing agricultural regime. In his report, *Agricultural Production Pattern Adjustment Programme for Punjab*, Dr S.S. Johl[18] has drawn up an ambitious plan to wean farmers away from cultivating paddy and wheat on 1 million hectares. If the Centre allows the state to implement this plan through central assistance, the consequent reduction in wheat production by 4.7 million tonnes and paddy by 3.4 million tonnes will mean a net saving of 14.7 billion cubic metres of water each year.

There is no doubt that a shift from the current cropping pattern will help the state curb the unintended and unanticipated trade in virtual water. Each tonne of wheat and paddy sent to the central food stocks entails a virtual transfer of 1,200 and 2,700 cubic metres of water respectively, i.e., the amount of water required to produce a ton of the harvest. The Punjab government has realised that in producing food for the nation, it is losing dearly in terms of its non-renewable natural resources for which there can hardly be any compensation.

The canalisation of rivers increased the water table in Punjab and it went up to 5 metres by 1979. Over exploitation since then and lining of canals made average water depth 28 metres again. What we gained in 120 years, we have lost in 30 years.

Inadequate Institutional Finance

Institutional finance accounts for 12 to 18 per cent of rural lending but covers only 40 per cent of the rural demand; the balance non-institutional loan carries an interest rate of 40 to 60 per cent. In exceptional cases, it goes up to 150 per cent according to a Reserve Bank panel.[19] The principle is that the poorer the person is, the smaller the landholding and hence the higher is risk of not recovering the debt.

---

[18]Dr S.S. Johl, *Agricultural Production Pattern Adjustment Programme for Punjab*, Report submitted to the Government of Punjab, 1986.

[19]Reserve Bank of India, *Report of the Technical Group Set up to Review Legislations on Money Lending*, 24 July 2007. Available online at http://www.rbi.org.in/scripts/ PublicationReportDetails.aspx?UrlPage=&ID=513

Village land is rapidly passing from the hands of the farmers to the *arthiyas* and moneylenders. Men whose forefathers tilled the same fields for generations are now forced to sell their land, often to settle 'debts' that may be illegal, entirely fraudulent or even non-existent. To rural Punjab, land is livelihood and more than that, it is life itself. Thousands of cases can be sited from the recent past wherein debt-trapped farmers and farm labourers, reduced to penury and threatened with the loss of their land, have committed suicide. Their number is rising at an alarming rate.

Subsidy to Farmers

According to Sompal, the former Union Minister for Agriculture, 'the total volume for subsidies available to farmers in India is just 10 per cent which is peanuts compared to what is available to the agriculturally and industrially advanced countries—60 per cent of the farmers received no subsidies in any form'.[20]

On an aggregate, rich countries spend around US$ 311 billion subsidising their farmers, against US$ 52 billion they give in aid to all countries. In other words, if the rich countries stopped all aid but removed farm subsidies and allowed developing and underdeveloped countries to export their farm products, the latter would benefit six times more.

It is not realised that the cheap foreign grain imports are the results of direct subsidy being given by those governments to their farmers.

Lack of Crop Insurance Cover

Businesses and industries enjoy full insurance cover, while farmers face all the risks of bad weather, pest attacks and unforeseen disasters without insurance. Farmers in other countries can insure their crops; no such cover exists for Indian farmers.

Occasionally the government doles out relief in worst-affected areas as a 'favour' but the relief sums are paltry, and come nowhere near the actual loss. The relief sum is first given to government departments rather than individuals and the money is received long

---

[20]Taken from an interview of Shri Sompal by P.P.S. Gill, 'Terms of Trade Are against Farmers', *The Tribune*, 7 December 1998.

after the crisis. In July 2004, farmers along the Ghaggar lost their crops due to floods. Relief was announced but as of February 2005, it had not come. In mid-February 2006 heavy rains destroyed crops on 9,000 hectares; a total of 2.3 per cent of all wheat under cultivation was badly affected.

For half a century, the government has dangled the promise of crop insurance, but done nothing. Unless farmers themselves get substantial relief well in time, they suffer grave financial losses which ultimately lead to suicides. Till such time as farmers have insurance, ad hoc relief to farmers should be increased to ₹20,000 per acre for wheat and paddy.

Inadequate Agricultural Research and Extension Effort

Before the Green Revolution of the 1960s, the practice in Punjab was to allow part of the land to 'rest' each year. A third of the land was left fallow. This land was also available for grazing and thereby was enriched by the manure of cattle. The Government of India, in collaboration with the Ford Foundation, an American organisation, pushed the Green Revolution onto Punjab. Punjab lost many of its traditional drought and blight-resistant varieties of wheat and gained a strong appetite for chemical fertilisers, pesticides and irrigation water. Crops such as pulses and oil seed were substantially replaced by water-guzzling paddy. The central government encouraged this by keeping a support price for wheat and rice but none for other crops. This coincided with the diversion of Punjab's river water to Rajasthan.

Higher education and research are strongly dependent on central grants. The Centre provides research money for HYV seeds but ignores the need for improving grain quality. Punjab was pushed to become the breadbasket of India and this policy helped so long as international grain prices ruled higher than domestic prices. The increase in domestic production and fall in international prices resulted in greater consumer selectivity in grain purchase. Price wise, Indian wheat is no longer competitive. It is therefore necessary to increase the allocation of funds to projects aimed at improving the grain quality and until this is achieved, the MSP must be retained at a remunerative level. Just as the British turned the Chinese into drug addicts at the beginning of the 19th century so that they would have a market for opium, the Centre, influenced by foreign interests, has ensured Punjab's addiction to growing wheat and rice.

Investment in Agriculture

There is a consistent decline in government expenditure on agriculture as depicted in the figures in Table 7.2. This has been a major factor in precipitating Punjab's agrarian crises. This crisis has been accentuated by the impact of globalisation. Farmers' suicides in India in general and Punjab in particular, both show this causal linkage.

The states have no independent power to contract loans, obtain foreign aid or draw on deposits with nationalised banking institutions. Public sector investments are the lowest in Punjab; while Maharashtra receives ₹15,000 crore, Punjab receives ₹457 crore only. This means that Punjab receives only 0.6 per cent against Maharashtra's 18.72 per cent.

Forest Cover

As per India's National Forest Policy, 33 per cent of the total land area should be under the forest cover. However, only 19 per cent is what we have achieved in the past.

In the past 20 years, Punjab's area under forest has declined from 6 per cent (see Table 7.3) to 3.09 per cent as per the *Survey of India* (*The Indian Express*, 28 July 2009).

This too appears doubtful. The actual area under forest cover would most likely be between 1–1.5 per cent.

Denudation of the forest cover causes accelerated soil erosion, besides reducing percolation and increasing surface run-off. Consequently, the loss of the forest cover reduces ground-water recharge and results in devastating floods. It takes nature 1,000 years to create a mere two inches of valuable top soil, which can be lost in 2 hours of heavy rain in a deforested area. In its greed to extract more grain out of Punjab; the central government treats the state as a colony and not as a valuable part of India which needs to be preserved and nurtured. At the time of Partition in 1947, every village had its

Table 7.2: Plan outlays for the Centre, states and union territories

(₹ Crore)

(1)	Total plan outlay	Agriculture and allied sectors	Irrigation and flood control	Total (3) + (4)	Total plan outlay	Agriculture and allied sectors	Irrigation and flood control	Total (7) + (8)
		At current price				At 2001–02 Price		
	(2)	(3)	(4)	(5)	(6)	(7)	(8)	(9)
Eighth Plan (1992–97) outlay	434100.0	22467.2 (5.296)	32525.3 (7.596)	54992.5 (12.796)	853830.25	44194.2 (5.296)	63979.0 (7.596)	108173.2 (12.796)
Ninth Plan (1997–02) outlay	859200.0	42462.0 (4.596)	55420.0 (6.596)	97882.0 (11.496)	1094541.7	54092.7 (4.996)	70599.9 (6.596)	124692.7 (11.496)
Tenth Plan (2002–07) outlay	152569.0	58933.0 (3.946)	103315.0 (6.896)	162248.0 (10.896)	1525639.0	58933.0 (3.996)	103315.0 (6.896)	162248.0 (10.696)
Annual Plan 2002–03 (Actual)	210202.9	7655.1 (3.040)	11964.8 (5.740)	19620.3 (9.376)	202333.92	7368.5 (3.096)	11516.9 (5.796)	18885.8 (9.396)
Annual Plan 2003–04 (Actual)	224827.0	8776.0 (3.540)	12800.3 (5.790)	21676.3 (9.090)	208407.1	8138.6 (3.996)	11963.3 (5.796)	20101.0 (9.096)
Annual Plan 2004–05 (Actual)	263665.2	10862.6 (4.296)	18024.5 (7.296)	28887.1 (11.976)	334370.1	9740.8 (4.296)	16004.2 (7.296)	26644.0 11.496

Annual Plan 2005–06 (RE)	351629.5	13439.8 (3.846)	25007.0 (7.196)	38446.8 (11.296)	299132.7	11433.3 (3.896)	21273.6 (7.96)	32706.8 (11.296)
Annual Plan 2006–07	441285.2	16162.8 (3.790)	33189.4 (7.590)	49352.2 (11.296)	354941.2	13000.3 (3.796)	16695.4 (7.596)	39695.7 (11.296)
Total Tenth Plan expenditure	1481616.2	56886.3 (3.8296)	102086.00 (6.8496)	150082.7 (10.6796)	1200183.08	40681.5 (3.8296)	22353.3 (6.8496)	138035.2 (10.6796)
Tenth Plan expenditure as % of outlay	97.896	96.796	98.196	25.196	91.296	24.396	25.596	25.196

Source: *Economic Survey, 2006–07.*
Note: Figures in brackets indicate 9 per cent of total plan outlays.

Table 7.3:   Forest cover and per capita forest area

Region/country	Forest cover to land area (%)	Per capita forest (hectares)
World	26.6	0.64
Asia	16.4	0.10
Africa	17.7	0.70
Europe	41.3	1.30
USA	23.2	0.80
Japan	66.8	0.20
India	19.0	0.06
Punjab	6.0	0.01

Source: Data based upon *State of World Forest*, Report by World Food and Agriculture Organization, 1999.

own forest land reserved for grazing. These village forests have all disappeared.

In December 1854, before transferring land to the white American Government, Seattle, the chief of a native American tribe spoke words that have since become famous. He said:

> Every part of the earth is sacred to my people. Every shining pine needle, every sandy shore, every mist in the dark woods, every meadow, every humming insect…We are part of the earth and it is part of us. Love it as we have loved it. Care for it, as we have cared for it. Hold in your mind the memory of the land as it is when you receive it. Preserve the land for all children, and love it, as God loves us. As we are part of the land, you too are part of the land. This earth is precious to us. It is also precious to you.[21]

Leaving a certain amount of land to lie fallow each year is a practice well known to the western farmers. It enables the land to recharge. This was also done in Punjab before the advent of chemical fertilisers. This practice should be revived. This can be done by paying an adequate non-cultivation subsidy to the farmers as is the practice abroad. The fallow land would be useful as grazing ground for sheep and cattle.

---

[21]Chief Seattle (1790–1866) was the leader of the Suquamish tribe of American Indians. He signed the Port Elliott Treaty with the United States in 1855 and along with it he made a statement. The quotation provided here is part of the same.

Food Grain Procurement, Storage and Transport

Agriculture is an industry for Punjab and procurement, transportation and warehousing of its food grains are its subsidiary
interests (see Tables 7.4 and 7.5). The benefit of these three
should be retained exclusively for the state. Agricultural profit for
the state can be enhanced by making favourable changes in these
three basic areas.

While it would not be possible to bypass price control, the central
government should be persuaded to transfer procurement to the
producer states. The Centre should function only as an allocating
agency and send timely allocations, failing which it should pay costs
incurred for delayed dispatches.

The procurement agencies of the state are paid handling charges.
To the extent that the procurement increases, the state will benefit.
This in turn will increase employment. The marketable food grain
surplus in Punjab was 178.47 lakh million tons for 2008–09. The state
has 192.45 lakh ton warehousing capacity which is inclusive of open
air storage. The capacity ratio should be 1:1.6. Because of the shortfall
in capacity, a large quantity of food grain is being stacked on open
plinths, with consequent heavy loses.

The finance for building additional warehousing capacity is
available from the World Bank. The World Bank advances loans
in the same way that ordinary banks extend loans to the FCI and
the Punjab Warehousing Corporation for building warehouses.
This will generate both temporary and recurring employment on a
large scale.

It is suggested that the producing states should prevail upon the
Centre to hold the buffer stock in producer states to give the benefit
of storage to these states. India's road network is well developed and

Table 7.4:   Procurement of wheat and paddy from Punjab

2008–09	(in million tonnes)
Wheat	9.939
Paddy	7.908
Total food grains	17.847

Source: Figures collated from various press clippings and reviewed by Dr H.S. Shergill.

Table 7.5:   Punjab storage capacity (includes open storage)

2005–06	(in lakh tonnes)
FCI	71.05
Food and Civil Supplies	07.69
Punjab Civil Supplies Corp.	20.91
Markfed	36.1
State Warehousing Corp.	36.06
Central Warehousing Corp.	07.00
Mandi Board	00.54
Agro-Industries Corp.	13.1
Total storage capacity	192.45

Source: Figures collated from various press clippings and reviewed by Dr H.S. Shergill.
Notes: This capacity is inclusive of open plinths.
     Number of trucks registered in Punjab as of 31 March 2006: 1,07,534.

it would be possible to move stocks to any consuming area within the country in 5 days.

During 2009, 178.47 lakh metric tonnes of grain were sent out by rail. If this quantity was to be shifted by road, it would require 37,000 trucks at an average of 40 trips per year carrying a load of 12 metric tonnes per truck. The number of registered trucks in Punjab as on 31 March 2006 was 1,07,534. This would represent 40 per cent utilisation of the state's fleet of trucks.

Amend the Transport Policy

To take full advantage for the state for this massive transportation of food grains, the transport policy would have to be amended. The Centre should be requested to:

1. exempt food grains from National Permit,
2. transport food grains through departmental fleet,
3. set up a goods transport body on the Jammu and Kashmir, Manipur, Nagaland, etc. pattern,
4. set up joint sector goods transport, and
5. encourage cartage through consumer cooperatives.

In case Punjab is able to persuade the Centre to raise India's buffer reserve, the advantages in terms of employment and cash flow can be mind boggling. Both storage and trans-shipment of food grains are post sowing and harvesting operations. This would provide off-season employment in the rural sector on a mass scale and thereby save many from committing suicide due to lack of alternate employment. Employment would be generated directly through construction activity, transportation, loading and unloading, *chowkidari* and support services.

Rail tariff is cheaper than road transport for cartage of food grains but when 6 per cent storage and transport loss, as permitted by FCI, and 4 per cent additional handling charges are factored in, the trans-shipment costs become even.

While the transportation cost is the same, transportation by road has the following advantages:

1. Warehouse to warehouse delivery.
2. Shorter transportation time.
3. National saving due to less pilferage and loss while in transit.

Many parts of the Northeast, interior Orissa and Madhya Pradesh are not directly served by rail. For such areas, dispatches by road are more economical and would generate off-season rural employment in the producer areas.

To bring in this kind of money, would substantially contribute to Punjab's prosperity and reduce rural unemployment.

Although trade in agricultural commodities was shifted from the State List to the Concurrent List, the suggested steps do not infringe its application. The greater danger is that the Centre, at a later date, might usurp agriculture itself and place it in the Concurrent List.

For the sake of value addition and generation of rural employment, Punjab should consider giving permission for marketing of paddy, only after shelling, that is in the form of rice. This would reduce transportation cost on the one hand and facilitate utilisation of rice bran for other purposes, in addition to generating substantial employment. Similarly wheat, unless it is being exported abroad, should also be sent to other states in the form of flour. Processed sugar should be sent out of the state rather than raw sugar cane.

Neglect of Rural India

Former Minister of Agriculture, Sompal spelt out the discrimination against rural India which has over 70 per cent of India's population. He mentioned that rural areas get only 26 per cent of the total electricity available, 11 per cent of housing loans, 18 per cent of higher and technical education, and 30 per cent of the schools. 'If all this is not discrimination, what else will you call it?'[22] he asked. Primary education in Punjab has collapsed due to low funding and in the universities the percentage of rural students now ranges between 2 to 6 per cent.

P. Sainath writes:

> It is not that inequality is new or unknown to us. What makes the past 15 years different is the ruthlessness with which it has been engineered, the cynicism with which it has been constructed and the scale on which it now exits. And that's at all levels, even at the top.[23]

He adds that, 'In one estimate, over 85 per cent of rural households are either landless, submarginal, marginal or small farmers. Nothing has happened in 15 years that has changed that situation for the better. Much has happened to make it a lot worse.'

Failure to Provide Education

The area of Punjab most affected by suicides is the area that has the lowest literacy rates. Literacy figures for the Lehra block is 29 per cent, for the Andana block is 28 per cent and for the Budhlada block is 28 per cent, as per the previous Census of India. Latest figures only show a continuation of the same trend. There has been no improvement.

The allocation for education in the central budget is made in the ratio of 70 per cent to higher education and 30 per cent to primary education. This has resulted in the lowering of standard in primary education and reducing the base of education at the rural level.

At the national level, Punjab is considered one of the most prosperous and developed states, but it is also a fact that around

---

[22]Taken from an interview of Shri Sompal by P.P.P. Gill, 'Terms of Trade Are against Farmers', *The Tribune*, 7 December 1998.

[23]P. Sainath, 'The Decade of Our Discontent', *The Hindu*, 9 August 2007.

80 per cent of the farmers are mired in debt and poverty. Rural students are barely able to pay the tuition fees, not to mention the cost of board and lodging, with the result that they are unable to attend colleges and universities which are located in distant towns.

The gap between the rural and urban is widening. This is clearly seen in college and university enrollment figures. Rural students account for a very low percentage of enrollments in institutions of higher education. The gap is so pronounced that it needs immediate correction.

Dr Ranjit Singh Ghuman, Sukhwinder Singh and Jaswinder Singh Brar of Punjabi University studied university enrollment in Punjab to find out how many students hailed from rural backgrounds. They found that only 4.07 per cent of the students came from rural areas. Of the total 22,360 students enrolled in Punjabi University, Panjab University, Punjab Agricultural University, Guru Nanak Dev University and their regional campuses, only 911 were rural students. The number of rural sudents is the least at Panjab University, a mere 2.20 per cent.

The study noted that there has been widespread exclusion of rural students in higher education in, particularly professional education, in the past two decades. The reason behind this may be attributed to collapse of school education in rural Punjab, admission through entrance tests, costly education in the private schools, the gap in rural–urban amenities and awareness, the information gap and the lack of guidance and coaching of the poor rural students. The study covered students who had passed their matriculation or senior secondary—plus two—examinations from rural schools, situated anywhere in the state.

Low literacy means severely restricted life opportunities. As farming income nosedives, the people of these areas are unequipped to take up any other livelihood except labour. For years, 70 per cent of central grant for education has gone to colleges and universities.

Its time to change the policy and turn that 70 per cent grant to primary education. The 2004 Union Budget provided a 2 per cent additional cess for education but that is not enough and the funds generated by this cess too are likely to go to higher education rather than where it is most needed.

Defence and Paramilitary Recruitments

Punjab is a border state with a 553-kilometre border with Pakistan. This stretch of land is densely populated. It is over this land that for centuries invaders have marched and attacked. Because of constant exposure to invasion, military service has long been a preferred occupation for the men of Punjab. At the time of Partition of India, Punjab's representation in the Indian Army was 27 per cent. This recruitment was almost entirely from the rural sector. After 1984, the criteria of recruitment by the Indian Army was changed from merit to proportionate representation, thereby lowering Punjab's share of recruitment to 2.5 per cent.

Assessing the strength of the Indian Army to be 15 lakh approximately and the retirement rate being 8 per cent per annum, annual recruitment would be around 1,20,000, out of which 32,400 was Punjab's share. This has now dropped to about 3,000. This represents a drop of almost 30,000 jobs per annum.

After Independence, the threat perception to India was from Pakistan and China. This threat has decreased significantly but the threat from internal disturbances has increased substantially due to disproportionate distribution of national wealth. The strength of the Indian paramilitary force is being increasing rapidly and today numerically it would be at par with the Indian Army. The total loss to Punjab in recruitment of these two forces is around 60,000 jobs per annum.

To underplay reduction of Punjabis in the Indian armed forces, the excuse now given out is that even 2 per cent recruitment is not possible from Punjab because applicants from this area are not medically fit to join the army. This is strange because Punjab has always been much ahead of other states of India in the field of sports, which indicates good health. The Indian Navy continues to prefer recruitment from the coastal areas.

The Indo–Pak border with Pakistan is an international danger zone and the danger is chiefly to Punjab. It is Punjab that has suffered most during the Indo–Pak wars. Punjabis have a vital interest in insuring the safety of its land and its people. The drastic reduction in recruitment to the defence services and the paramilitary, without providing alternate avenues for employment, has been a major factor in the spurt in rural suicides. The Union Defence Ministry should increase recruitment from the strategic, densely populated forward areas of Punjab.

Sealed Border

Before Partition, Punjab's progress owed much to lying along the trade route to West and Central Asia, both by land and by river. Punjab's cities flourished thanks to their location on this trade route. Its five rivers flowed into the Indus carrying merchandise to the port at Karachi for onward shipment. Both Punjab and Bengal suffered the tragedy of Partition; Bengal was not deprived of its port or the trade that flowed through it, but Punjab lost a significant economic resource. Reopening the border with Pakistan would not only allow Punjab to benefit from favourable terms for agricultural products, it would also benefit Himachal Pradesh, Jammu and Kashmir, and Haryana in particular and India in general.

The borders are sacrosanct and a number of wars have taken place even on the slightest attempt to change a border. Strangely, the Indian government has shown a total lack of sensitivity on this issue. In 1947, they agreed to divide India on communal lines. They gave thousands of square kilometres to China in the Aksai Chin area on the tenuous plea that the area has nothing but 'barren stones'. Even on the Punjab border, in 1988, in a very cavalier manner, the government sought to build a security belt along the Punjab border which was 5 kilometres short of the actual border. They were prepared to deprive farmers living there of thousands of square kilometres of highly productive land. This area is densely populated. Large-scale resettlement of population was ordained; hence the scheme would have required Parliament to apply Section 249.

As a first step in this direction, Article 249 was steered through the Rajya Sabha; this allowed Parliament to make laws on certain matters enumerated in the State List. These were law and order, police and jails. So long as the resolution was in force, the Parliament had the power to legislate on these subjects within the state or part of the state and the state executive was bound to carry out and implement the laws on these subjects.

It looked as though the union government was unwilling to depend on the population along the international border as a potential source of effective defence as they did in 1965 and 1971. Was it that the government thought that the local population had reason to be alienated? Was anything done to ensure that the people were not alienated? Can any amount of force compensate the loss of people's support?

The implications of the scheme were conveyed to all members of the Indian Parliament by a group of concerned citizens—a former Additional Advocate General, Mohinderjit Singh Sethi; Major General G.S. Kalkat; and in the field of academics, an economic geography specialist, Dr G.B. Singh of Punjabi University and linguist, Dr S. Prakasam of Punjabi University—who were led by I.S. Jaijee, at that time a Member of the Punjab Legislative Assembly.

The advice tendered to the MPs—based on an on-the-ground survey of the entire border area—cautioned that the scheme, if implemented, would carry a huge human and economic cost. Based on the assessment of these concerned citizens, the evacuation of a 5-kilometre broadband would have entailed infrastructural loss—at prices prevailing at that time—of not less than ₹5,000 crore approximately and a loss of the annual produce of the land amounting to ₹160 crore, sufficient to feed 5 million hungry Indians. It was pointed out to the government that while developed countries such as Japan were putting up airports, residential suburbs and commercial districts on the sea, India was thinking of implementing a scorched earth policy involving its most productive land. Creation of a belt would have entailed a displacement of more than 4 lakh people.

Within Punjab, the districts of Gurdaspur, Amritsar and Ferozepur fall on the Indo–Pak border. The Gurdaspur section runs mostly along the river Ravi and remaining small stretches are covered by ditch and bund defences. The Amritsar district is entirely protected by ditch and bund defences. The Ferozepur district has a large river border (the Sutlej) and the rest is covered by ditch and bund defences. In addition, sensitive areas are mined. A total of only about 60 kilometres, in stretches of various lengths, on the Indo–Pak border are vulnerable to smuggling and infiltration.

The report suggested an alternative arrangement for the vulnerable stretches, namely, a double fence enclosing parallel and lateral roads, watchtowers at suitable distances equipped with search lights and electronic devices, and aerial surveillance by light planes equipped with up-to-date electronic sensing devices. It was also suggested that the vegetation should be controlled (elimination of elephant grass) and plantation restricted to low-height crops. Compensation to farmers of border villages for crop-loss was recommended.

Normally, it is the weaker man who erects barriers against his stronger rival. It would have been logical if Pakistan had been keen to fence the border to protect itself from India. Instead, India has shown greater anxiety over porous borders and a greater interest in stopping all cross-border contact. India's backtracking on pre-Independence promises made to the states has resulted in simmering internal discontent. India's leaders have long resorted to stirring up fear of 'foreign hands' but even a political simpleton can see that this was a ploy to manipulate the people. In fact, it is the long-building frustration of India's own people that political leaders fear.

The research and suggestions offered received wide appreciation and assurance of support from a number of influential MPs, including S.B. Chavan, P. Upendra, Buta Singh and a number of other MPs. The study effectively thwarted the 5-kilometre security belt scheme. Ultimately, the government's border scheme was not adopted. The depth of the belt was reduced from 5 kilometres to 1.5 kilometres and the border was fenced. Unfortunately, even this fence left quite a few chunks of territory on the wrong side and farmers were not compensated in any way.

The sealing of the border with Pakistan vis-à-vis Punjab and Kashmir is the direct fallout of reneging on promised regional autonomy to the states at the time of transfer of power. It was feared that Pakistani contact might kindle fires of discontent among disgruntled neighbours. Punjab has always been regarded as especially dangerous. Every one of Punjab's demands was seen as a 'camel's nose' in the tent of full autonomy—yield a little and not only would demands multiply but other states would follow Punjab's example.

What however was not visualised at that time was the cost to the farmers of cross-border trade.

In the decades after Partition, a regular trade of this nature continued and took care of surpluses and deficits on either side. The character of this trade was entirely people-to-people. Similar trade even now can be seen on the Gujarat–Pakistan borders,

Rajasthan–Pakistan border and the borders of India's Himalayan states with China, Nepal, Bhutan and Burma. In the case of Bhutan and China, only a small number of trade routes exist due to the mountainous terrain and these have always been active since time immemorial. In the case of Rajasthan and Gujarat, desert conditions and sparse population have limited open trade to a few points. In Kashmir, hostilities around the valley precluded westward trade; moreover the sparsely populated mountainous border provides no encouragement.

Punjab's case is entirely different. To begin with East and West Punjab were one unit until only 60 years ago. The area is a geographic unity with no natural dividing features. The natural direction of Punjab's economic gaze is westward—extending to Karachi in the south and Afghanistan in the north the gateway to even more distant trade. Punjab's rivers flow into the Arabian Sea; although the fabled Saraswati has a mythical *sangam* at Prayag (Allahabad), recent satellite imagery indicates that the now extinct river actually debouched in the Arabian Sea. Punjab, both east and west, is heavily populated and highly productive—both factors being strong incentives to trade. Westward trade built many cities of Punjab, notably Amritsar, Jalandhar, Ferozepur, Ludhiana, Gurdaspur, Ambala, Karnal and Panipat.

In villages directly on the border it was common for farmers to cultivate land on the West Punjab side of the international border. It was said that they reaped from under the very guns of the Pakistan Army. Such cultivation demonstrated the Punjabi farmers' confidence and high morale and indirectly India's strength. Just as farmers cultivated across the border, so too did farmers and local merchants buy and sell from their counterparts across the Radcliffe Line. From an official point of view, it was smuggling, but viewed simply as an economic operation, the trade was logical and profitable for both parties.

Facts about the Fence

At present, the electrified fence is between 100 metres to 1.5 kilometres within the border. Put up in the late 1980s to halt terrorist infiltration into Punjab, the fence is located within the Indian territory.

Table 7.6:   State-wise progress of border fencing

State	Total length of border	Length sanctioned for fencing	Fenced so far	Remaining to be fenced
Punjab	534	461	462.45	—
Rajasthan	1,037	1,056	48.27	—
Jammu	210	180	184.59	—
Gujarat	508	310	217	93
Total	2,308	2,007.63	1,912.31	93

Source: Ministry of Home Affairs, *Fencing and Floodlighting of Borders*. Available at mha.nic.in/pdfs/Fencing.pdf
Note: Figures in kilometre.

Due to the zigzag border, authorities had decided to keep the fence in a straight line at most places to make monitoring easier (see Table 7.6).

This led to vast tracts of agricultural land being left in the bulges on the other side of the fence. Over 15,000 acres of land came on the other side an area equal to a district of Punjab has been engulfed in this exercise. What pinches the fence-affected farmers the most is the fact that the Punjab government compensated farmers for their land acquired, for setting up a cargo station at Attari border at a rate of ₹3.5 million per acre while they have not got even a single paisa for their land.

The fence displaced thousands of border farmers whose agricultural land fell on the other side of the fence. The farmers were allowed restricted entry to their own fields by the Border Security Force (BSF) after thorough, and at times humiliating, frisking at the fence gates to ensure that no weapons, drugs and humans are smuggled in.

Agricultural activity for these farmers is restricted to just a few hours between 10 a.m. and 4 p.m. In actual practice, permission is given for a mere 2 to 3 hours. It has placed the farmers whose land falls on the far side at a great disadvantage and put them to enormous amount of inconvenience, because the gates are widely separated. The first and foremost problem is one of the watering, because of the problem of keeping the water channels functional, especially across the width of the fence and watering at night.

Pakistanis often drive their cattle at night to feed on the crop and there is no way a farmer can guard his fields. There is considerable amount of harassment in getting through these gates. Often the guards must be bribed and the fixed timings of the opening and closing of the gate harasses the farmers.

With Indian bunkers behind the fence, Pakistan, at most places, has moved its bunkers right on the border and these intimidate the farmers working in their fields, who work right under the shadow of Pakistani machineguns.

For the past decade, India has produced surplus food grain. Logically, surplus food grain would flow to grain-hungry regions such as Baluchistan, Afghanistan and the Central Asian republics to the benefit of India—particularly the farmers of North India in general and Punjab in particular. Finding a market for our food grain is now, more than ever, of utmost importance to us.

International oil companies are presently chalking out plans to bring oil out of Central Asia via a trans-national high-tech pipeline. The talk presently centres on pipelines from Baku in Azerbaijan to Basra in Iraq and thence to Bandar Abbas in Iran and eastward under the Arabian Sea to Karachi and Okha in Gujarat. This is an expensive venture that could be avoided if India and Pakistan were on sufficiently good terms to allow an overland pipeline eastward out of Central Asia.

Circumstances have altered radically in the past 20 years and fresh thinking is required. With the necessity of international trade and tremendous advances in communication, it no longer makes any sense to advocate any sort of 'curtain'—iron, bamboo or even silk—around one's country. The aim must be the normalisation of the Indo–Pak border. The rate of suicide all along the Indo–Pak border is very high. The bigger loser on account of the sealed border has been India as it has lost its geographical advantage for trade not only with Pakistan but also with Central Asia and beyond. This is the era of globalisation and opening up of world trade. Bilateral trade with a neighbour, hostilities notwithstanding, has become inevitable.

Cross-border Trade

Mutual trade between India and Pakistan has been steadily increasing since 1989 in spite of the fact that Pakistan is yet to accord Most

Favoured Nation (MFN) status to India. In addition to official trade, an unofficial trade worth about US$ 2,000 million annually is taking place between them. This trade is routed through third countries like Dubai, Singapore, Hong Kong, Afghanistan and the Gulf-States. This tortuous route costs both India and Pakistan dearly in terms of foreign exchange outflows as well as third party trading commission. In addition, goods worth more than US$ 1,000 million are smuggled across the borders every year. Through illegal channels India exports stainless steel utensils, cosmetics, alcoholic beverages, ayurvedic medicines, cotton items and confectionary items (see Table 7.7).

These figures do not include trade volume of India's trans-Pakistan trade with Central Asia and beyond or fringe benefits such as growth of our goods transport sector or diversification of agriculture to cater to cross-border demand.

The value of India's trade with China tripled within 3 years by normalising relations. Normalising relations with Pakistan would see a similar rise in trade and benefit both the countries. Both Pakistan and India will get a revenue boost through custom duties if unofficial trade is channelled through official routes.

Failure to take this step makes no economic sense. Why should trade between the two neighbouring countries be mainly via sea, which is much more costly and time-consuming? A direct flight from Delhi to Lahore takes 2 hours and costs ₹6,675 on business class, but via Dubai it takes all day and costs ₹57,250. Only 50 kilometres separate Amritsar and Lahore. Pakistan can get raw material and other items 25 to 40 per cent cheaper if imported from India.

Recently, a cotton mill in Karachi refurbished its machinery. Every component was provided by a major textile machinery manufacturer

Table 7.7: Indo–Pak trade

Year	Trade volume (in US$ million)
1989–90	63
2001–02	204
2005–06	859
2006–07	1,000

Source: Ranjit Singh Ghuman, 'Trade across Borders: India and Pakistan Must Keep the Momentum Going', *Times of India*, 7 October 2007.

in south India. It was first shipped to Dubai and then to Karachi. The Pakistani paid more, but the Indian company got nothing. Apollo is the most popular tyre brand in Pakistan. Apollo tyres are smuggled into Pakistan via Dubai, Singapore and the Punjab border, as the import duties are around 46 per cent.

India produces engineering goods and there is a strong market for these in Pakistan. Punjab has very heavy-rolling capacity for steel items. Whenever trade with Pakistan opened up in a limited manner, rolled products were exported to Pakistan. Moreover, Punjab's entrepreneurs have developed a range of products in the automobile sector that are exported to advanced countries in large volumes. Likewise both countries will benefit by trade in agricultural products. If Punjab's farmers were allowed to sell their cotton and wheat to Pakistani traders, they would be getting ₹200 to 300 more per quintal than in the domestic market. Exchange of perishable goods like fruits and vegetables would also expand.

Presently cotton from East Punjab is exported to Karachi (Pakistan) via Kandla port, which is very expensive due to heavy transportation charges. It would make better sense to open the Indo–Pak border right from Dera Baba Nanak in Gurdaspur up to Hussainiwala in Ferozepur and set up duty-free ports along the Indo–Pak border.

The Punjab government should prevail upon the Union government to ensure that in any trade dialogue with Pakistan, Punjab entrepreneurs are closely involved and should have direct dialogue with their counterparts in Pakistan.

Pakistan can export power to India. Private producers have created spare capacity in the power sector in Pakistan. There have been disputes on the power rates although rates there are much cheaper than in India. Punjab and the other northern states can benefit on this front as well, as there is an acute shortage of electricity in Punjab, Haryana and Rajasthan.

Kashmir Chief Minister Mufti Mohammad Sayeed was right in demanding a soft border. Apart from trade, he has asked why Kashmiris cannot freely meet their fellows across the border if people in Gujarat and Rajasthan can do so. Punjab has an even greater stake in open borders. The concept of 'soft border' with Pakistan, envisages opening of trade 'gates' at selected places to enable government-regulated trans-national trade. It does not take into consideration trade of the informal kind mentioned previously. To facilitate people-to-people trade, more trading points should be opened along the Indo–Pak border.

Fifty years of hostility have sapped both India and Pakistan. Instead of investing the countries' resources in development and improving quality of life for all citizens, mutual fear has driven India and Pakistan to three wars and the maintenance of huge standing armies in a state of perpetual war-readiness. If this huge sum alone were diverted to development and cross border trade resumed, India would be at least as prosperous as South Korea or Taiwan.

Let common sense dawn at last. The northwestern trade, apart from generating trade benefits, would have taken care of a sizeable number of surplus labour from farms, provided higher returns from the farms and the farm produce, and would have reduced costs due to availability of cheaper items like fruits, vegetables, etc.

The best way to normalise relations between India and Pakistan, and eliminate terrorism is to integrate the economies of the two countries, thereby boosting the economies and reducing defence expenditures for both. This would be part of a larger policy configuration founded on a free and willing association of autonomous states. The oft-repeated slogan of unity in diversity must become a fact before we can reach the vision set forth in Iqbal's wonderful verse: *Sare jahan se achcha, Hindustan hamara.*

Industrial Freight Corridor

India is in the process of creating special freight corridors to facili-tate transport of goods for export and import. The route of these corridors (Mumbai to Dadri) passes through Maharashtra, Madhya Pradesh, Uttar Pradesh, Rajasthan and Haryana. On the representation by Punjab, it has now been extended to Ludhiana via Chandigarh. Chandigarh is not Punjab. This appears to be a concession to the recently set up Baddi area of Himachal Pradesh as an industrial complex.

The Yamuna provides the boundary between the states of Haryana and Uttar Pradesh on the eastern sides. The proposed route runs close to the Yamuna from Delhi to Karnal which is a distance approximately of 150 kilometres. The land on the eastern side of this route from Delhi to Karnal is a narrow strip. The main trade would therefore come mainly from the western side.

This corridor is already developed and can handle increased cargo; investing money on improving it may give some increased benefits in the short term but developing an alternate route through the interior of both Haryana and Punjab would provide much higher returns and benefit a considerably larger section of the people. It would be much more inclusive (see Map 7.1). This map shows the present route (Delhi to Baddi in Himachal Pradesh) and suggests an alternative

Map 7.1: Freight corridor routes

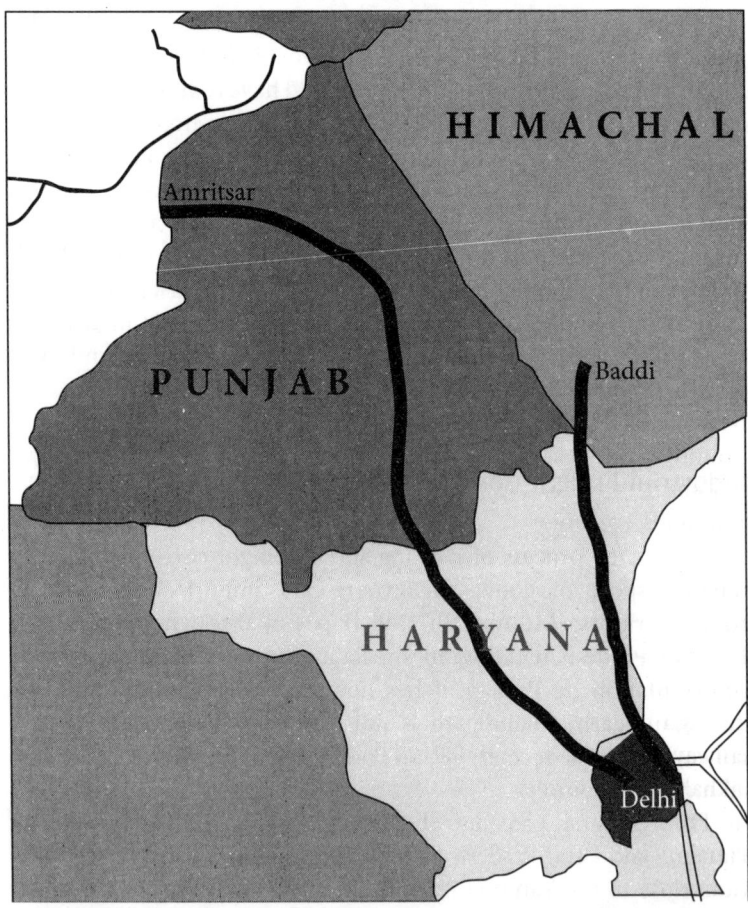

Source: *This map has been redrawn, not to scale, and does not depict authentic boundaries.*

route passing through the interiors of Haryana and Punjab (Delhi to Rohtak to Jind to Narwana to Tohana to Jakkal to Sunam to Dhuri to Malerkotla to Ludhiana to Jalandhar to Amritsar).

After the Partition of India in 1947, industrial development was denied to Punjab, first on the rationale that it would be vulnerable to the enemy and later on the consideration of retaining Punjab as the bread basket of India. Whatever industrial growth took place was mainly around the periphery of Delhi. On the division of Punjab in 1966, these industrial areas went to Haryana. Here again while the state of Haryana around Delhi and along the eastern end of the corridor benefited, the interior was left undeveloped.

The land along the corridor up to Karnal is highly fertile and should be reserved for the agriculture. On the other hand, the land along the proposed alternate corridor passes through central Haryana, an area that is semi-arid and water is highly saline and not suitable for agriculture. In the Punjab section of proposed route, the sub-soil water apart from being saline is receding. The average depletion rate is 30 to 40 centimetres and goes up to 1 metre in some places, conjuring images of impending desertification in the very near future. Another advantage in choosing an alternate corridor, as indicated in Map 7.1, is the cost of land acquisition which is much lower.

The alternate route would serve almost all the important towns in both the states of Punjab and Haryana. This would have a shorter distance. It would help to strengthen and extend industrial growth around these towns. This in turn would generate employment and absorb surplus farm labourers who are committing suicide in thousands.

The proposed alignment travels along the middle of both Haryana and Punjab, drawing maximum benefit of industry on both the sides. This travels along Delhi, Rohtak, Jind, Narwana, Tohana and Jakhal in Haryana. In Punjab, it goes through Sunam, Dhuri, Malerkotla, Ludhiana, Jalandhar and Amritsar. This passes through the middle of Haryana and Punjab and is well served by railway network and junctions at Rohtak, Jind, Narwana and Jakhal in Haryana which link Bhiwani, Panipat, Kaithal and Hissar. In Punjab, Dhuri, Ludhiana and Jalandhar junctions connect Patiala, Bhatinda, Khanna, Gobindgarh, Ferozepur and Jammu. Connectivity by road is good but it requires further strengthening. An eight-lane road from Delhi to Rohtak has already been taken up. The corridor needs to be extended up to Ludhiana and the single rail track needs to be upgraded to a double track. No major obstacles such as rivers fall in the way of this route.

This proposed corridor would serve as a backbone to the network of strategic defence roads along the Indo–Pak border. Since the time the corridor plan was thought out, our relations with the northwestern state of Pakistan have undergone a sea-change. Democratisation of Pakistan has made trading with our neighbour not only possible but desirable. In response to this opportunity the corridor project route needs to be revised. It makes sense to route the corridor so as to facilitate trade with Pakistan also. Amritsar to Lahore is a mere 50 kilometres and Hussainiwala border from Ferozepur is just as close. The alternate corridor proposed would provide access to all the three main trading routes to Pakistan, i.e. Hussainiwala, Attari and Dera Baba Nanak.

The concentration of industry along the Grand Trunk road in both Haryana and Punjab provides alternate employment to the displaced rural labour. Consequently, rural suicides along this belt are very few. It is therefore necessary to improve industrial growth in the interior of these two states. This would facilitate even development.

Punjab's Perpetual 'Debt' to the Centre

On his 22 March 2010 visit to Chandigarh, Union Finance Minister Pranab Mukherjee told the media that the Centre is not considering debt waiver for Punjab. The debt referred to was the debt incurred some 20 years ago to suppress militancy in Punjab. According to figures published in the media, this accounts for nearly ₹380 billion or more than half of Punjab's total ₹649.24 billion debt. The state pays an annual interest of ₹53.89 billion to the central government. The debt continues to mount and it is projected that the state's total debt by 2010–11 will rise to ₹71,086 crore—and in consequence interest will rise to ₹57.64 billion. The state's gross domestic product (GSDP) for 2006–07 stood at ₹911.48 billion. This means that Punjab is paying about 6.32 per cent of its GDP as interest.

According to former Punjab Finance Minister, Manpreet Singh Badal, during the 10 years of terrorism (1983–84 to 1993–94) revenue and tax collection was almost negligible but debts mounted on account of providing boarding, lodging, vehicles, fuel and ammunition to central forces called in to aid the state. He mentioned the sum as ₹67.72 billion. (This principal is nearly equal to the interest demanded in 2009–10 and is less than the interest due in 2010–11.)

The continuing existence of this debt puzzles the citizens of Punjab because they have clear memories of the visits of then Prime Minister Narasimha Rao in May 1991, then Prime Minister I.K. Gujral in May 1997, and present Prime Minister Manmohan Singh in September 2006. All these prime ministers announced that Punjab's debt had been waived. The only prime minister who did not make such an announcement was Atal Behari Vajpayee. Perhaps he thought that the debt had been waived previously and hence there was no need for him to waive it.

From the Finance Minister's statements it appears that the state of Punjab is not contesting the continued existence of this loan, but it ought to. Consider this: If a private individual challenged a debt in court, the court would inquire closely into the circumstances under which the loan was contracted. The court would ask to see the loan documents and from that the court would ascertain when was the loan was contracted, the terms and conditions of the loan and who signed the loan.

In Punjab's case, the loan was supposedly contracted when Punjab was under President's Rule, i.e. the Governor's rule. Governor's rule means that no elected representative of the citizens of the state is consulted or plays any role in governing the state. The governor is appointed by and functions as per the directions of the Centre.

So here is the sequence:

1. The Centre decides that Punjab needs a loan.
2. A loan is drawn up.
3. Only the governor is available to sign it.
4. No elected representative is consulted about it or is shown the loan document or has any opportunity to study the terms and conditions.

This is like a bank deciding that an individual needs a loan and an agent of the bank signs the loan committing the individual to repay. The individual has been given no chance to either accept or reject the loan; even his thumb impression has not been taken.

Unlawful and Discriminatory

The loan was purportedly taken to meet the expenses of suppressing militancy in Punjab. Punjab was affected by militancy for roughly a 10-year period from 1984 to 1994. Differences with the Centre were initially focused on the Centre's forcible appropriation of Punjab's river waters. Punjab Chief Minister Darbara Singh was coerced into signing the 1981 agreement that left Punjab with only 4.22 MAF of water. The state legislative assembly was bypassed. Thereafter the Centre announced that a canal would be built to link the Sutlej and the Yamuna. The farmers of Punjab, led by the SAD, rose up against this canal. As the conflict escalated, it was manoeuvred into the form of a religious and secessionist struggle. The fatal turning point came in June 1984 when the Centre sent the Indian Army into the Darbar Sahib, against the advice of the Punjab Governor, A. P. Pande. Pande was relieved of office shortly after the Army operation. The conflict was thus not of Punjab's making.

The central forces—Army, Central Reserve Police Force (CRPF) and other paramilitary organisations—were called in to suppress militancy in Punjab. The purpose of the Army is to defend India and maintain the integrity of the country. The Army is not a private security agency whose services are taken under contract. Part of every rupee collected from taxpayers goes to fund India's Defence Forces. The people of Punjab are paying their share toward maintaining the Army and other paramilitary services as a matter of course. Why should they be made to pay twice over—once as tax and once as loan repayment?

The northeastern states have been affected by militancy even before Independence and militant groups are active in these states even now. Kashmir has been affected by militancy from the 1990s and continues to be so. Since the 1990s Naxalites have spread over West Bengal, Orissa, Jharkhand, Chhattisgarh, Bihar, parts of Andhra Pradesh, Maharashtra and Uttar Pradesh. Central forces are deployed in all these states. Have central loans been advanced to these states for the purpose of suppressing militancy? If so, what is the quantum of these loans and the terms and conditions of the loans?

If Punjab has been committed to a central loan but not the other states, then the action is discriminatory. More than that, it is punitive since the citizens of Punjab are deprived of money that might otherwise be spent for the betterment of the state. For what offence are the citizens of Punjab being punished?

In the decade of militancy, Punjab was under stress and the state remains under stress now also. Rather than helping to lift up the state out of its difficulties, the Centre is bent on crushing Punjab further. For the sake of inclusive development, the Centre should pump in more, rather than take away from the state.

Punjab is an integral part of India. The Army and other central paramilitary organisations were deployed in Punjab to suppress militancy, prevent militants from ousting the established government and administration of the state and taking Punjab out of the Indian Union. In combating militancy in Punjab, the Army and the other forces were doing their duty and not rendering a paid service or favour to the state of Punjab.

As SAD Finance Minister Manpreet Badal pointed out in 2009: 'The Constitution provides that in case of any internal and external threat, the Central Government is bound to protect the state both physically and financially. But, in our case, the centre has abdicated from its constitutional obligation by not waiving off the debt.'[24]

To an Indian, every inch of the country is equally sacred. Remote and inhospitable areas should not be dismissed as 'barren stones'; likewise border states should not be denied on account of their vulnerability. In the case of Punjab, this state feeds the country. The people of this state should not be oppressed to the point where they escape through suicide.

Punjab's case should be that the purported loan to Punjab (documents of which may or may not exist) is illegal. As the Punjab Assembly formally repudiated all previous river waters agreements in 2004, likewise it may repudiate this purported loan.

CENTRAL SUPPORT TO SUICIDE PRONE STATES— EXCEPT PUNJAB

Incidence of farmers committing suicide in certain parts of India has been a matter of serious concern to the Centre. A rehabilitation package of ₹1,697.869 billion was worked out and the recipient states were identified as Andhra Pradesh, Maharashtra, Karnataka

[24]Manpreet Randhawa, 'Terrorism over but Punjab Continues to Reel under Heavy Debt', *Hindustan Times* (Chandigarh edition) 17 March 2010.

and Kerala. To this package was added ex-gratia assistance of ₹155 million, interest waver of ₹270.781 billion and restructuring and rescheduling loan of ₹905.112 billion. A special package was also given to Vidharbha. Punjab and Haryana were excluded.

Later a Dry Land package was given to the states that were doing dry-land farming. Punjab was again excluded on the ground that Punjab was not a dry-land area. The central government overlooked the cost of extraction of underground water in Punjab where the water table is sinking rapidly and sub-soil water is contaminated due to over use of fertilisers and pesticides.

In 2008 ₹710 billion was ear-marked as national debt waver. Though Punjab contributes approximately 50 per cent grain to the national pool, but of these billions, Punjab received a share of only 1.3 per cent.

As Punjab has all along been projected as the most progressive and well-off agriculturist state of the country, admission of rural suicides in Punjab would expose the central government, thereby compelling it to restructure its planning in favour of the agriculture sector. On the other hand, under political compulsions, the central government is resorting to denying central aid to Punjab's rural sector.

In spite of the phenomenon of rural suicides in Punjab being much graver than that observed in the southern states, the Punjab government is desperately trying to under-play rural suicides in Punjab as it is short on funds to provide financial support to the distressed families. This effort to hide farmer suicide is not only peculiar to Punjab alone. This is also observed in the other states and has been widely reported by the media.

8

Quantification of Suicides

INTRODUCTION

The farmers of Punjab are no strangers to hard times and in-security. Over the centuries, they have watched their crops perish in floods and droughts. In the past, devouring swarms of locusts left famine in their wake and marauding armies plundered and burned everything in their path. The village moneylender and the *seth*s of the *mandi*s have exploited the farmers ever since money was invented. Likewise, feudal lords bled the farmers for *begar* and taxes.

In ages past, the response to relentless oppression was often rebellion. Pushed to the edge, a villager might take to dacoity. Punjabi folklore is full of tales of heroic outlaws. In fact, many of these Punjabi Robin Hoods hailed from the area of this study. They were dacoits in the eyes of the Mughal authorities and later in the British colonial government, but heroes as far as the local people were concerned. One such outlaw hero was Jeona Maur. Maur is a small town about 30 kilometres from Lehragaga.

The patience and fortitude of Punjab's farmers is legendary. The overwhelming majority of villagers in Punjab are Sikhs. Sikhism is a strongly life-affirming faith. The 10 gurus of the Sikhs, particularly the revered last guru, Guru Gobind Singh, commanded the Sikhs to struggle against oppression rather than submit to it. Suicide is strongly disapproved in the Sikh religion and so is begging. Sikhism, and indeed the Punjabi ethos, abhors begging. The beggars seen in Punjab are generally 'professional beggars' from the eastern or southern states or from Rajasthan. Very few beggars hail from Punjab. When debt reduces a farmer to absolute penury, he is more likely to take his own life rather than beg.

CHANGING NATURE OF OPPRESSION

It is only in very recent years that India in general and rural Punjab in particular have begun to record accurate individualised data on births and deaths. Recording cause of death is still regarded as a tall order. Therefore it is not possible to declare as a substantiated fact that rural suicides in the decades before or immediately after Independence were negligible. However, one finds no reference to rural suicides in Punjab in either folktales or literature. The classic pre-Independence study of rural Punjab, namely *The Punjab Peasant in Prosperity and Debt*[1] by Sir Malcolm Darling (ICS), makes no mention of it. Darling made an in-depth study of the severe impact of the Great Depression of the 1930s on Punjab's farmers. If farmers had been committing suicides in response to their economic hardships, then he would certainly have taken note of it.

However, the nature of the oppression faced by rural Punjab has changed. Fleeing into the jungle or the desert does not help and certainly offers very limited scope for hitting back or effecting any change for the better. Twenty-first century oppression can be summed up as a deliberate rigging of the political structure to accelerate the growth of inequality. The motto is *Garib Hatao*—even as the media constantly celebrates India's growth rate, it's future as one of two mighty engines driving the world's economy (the other being China) and the consuming power of the Indian elite. It is true that India is richer today than it was a decade ago but this wealth has poured into the pockets of the very top segment of the population. Those at the lower end of the economic scale are worse off than ever before. For an overview of suicide-prone areas in

---

[1]Sir Malcolm Darling, *The Punjab Peasant in Prosperity and Debt* (Delhi: Manohar, 1977; reprint of the 4th ed., 1947, with a new introduction by C.J. Dewey).

Punjab as a whole, and especially in district Sangrur, see Map 8.1 and 8.2, respectively.

Development in British Punjab also translated into better education and more awareness of the outside world. Most of Punjab's outward migration has taken place from erstwhile British-Punjab

Map 8.1:   Punjab: Suicide-prone districts

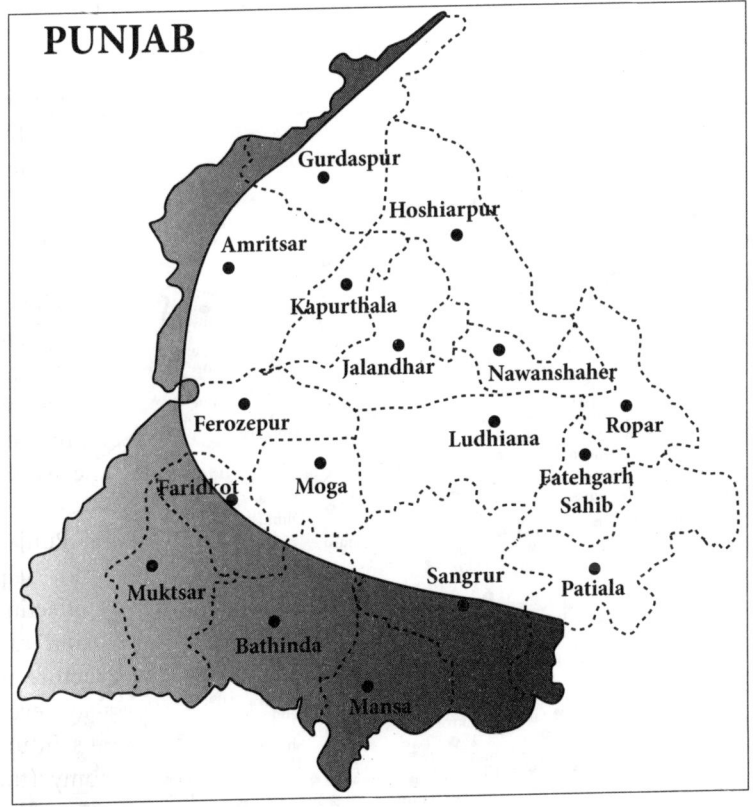

Source: Author.
Notes: (i) Dark area represents higher rate of suicide.
(ii) The rate of suicides is higher in the area covered by the former princely states of Punjab and along the Indo–Pak border. The development strategies of the Punjab government since the merger of Patiala and PEPSU into Punjab was to strengthen the infrastructure laid out by the British government in Punjab. The development has therefore been much more in the erstwhile British-Punjab region. This has helped to provide alternate employment to residents of rural communities in that region resulting in fewer suicides.

areas. This migration has helped to take care of surplus agro-labour and inflow of remittances from abroad softened agro-desperation.

While there may be a security threat along the Indo–Pak border, there is no threat whatsoever to deter equal development of both Punjab and Haryana along their internal borders.

Border areas, especially those on the Indo–Pak border, suffer from isolation and from one-sided development. The state's investment on infrastructure development should take this factor into account.

Map 8.2: Lehragaga block, district Sangrur: Number of suicides documented in each village

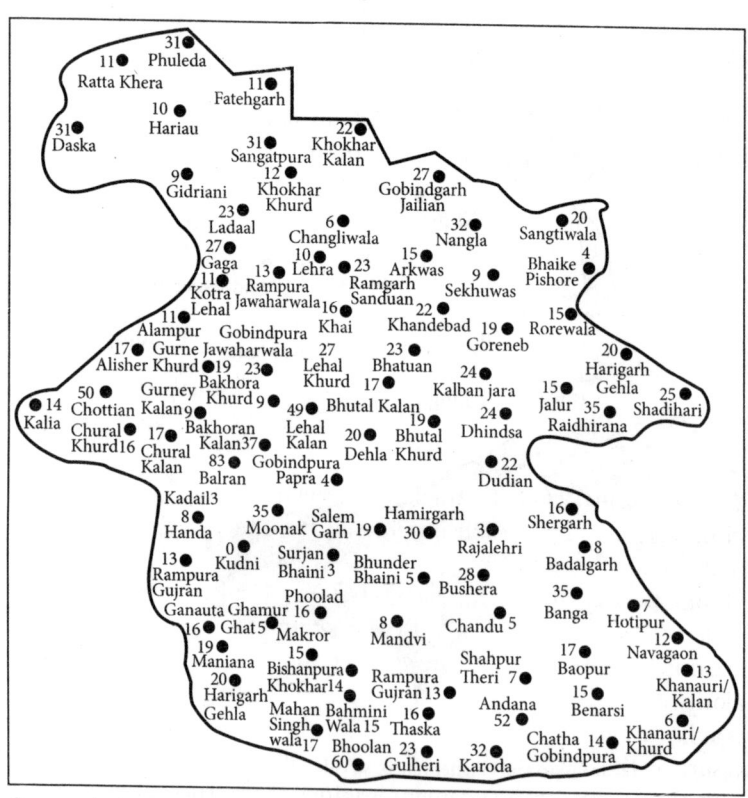

Source: Author.

Notes: The data represented in the map is from 1998 to 2008. In the two subdivisions of Lehra and Moonak, it is observed that the villages along the Haryana border have a higher rate of suicide. The Haryana villages along the Punjab border also suffer more suicides than those more within the border.

In terms of rural Punjab, more than 85 per cent of rural families are either landless or barely surviving on 4 acres or less. Their survival is becoming more precarious every year—not because they are lazy or unskilled farmers but because the government is bent on making it impossible for farmers to farm. Can the farmer resist? All the levers of his economy are in the hands of the government—input costs, crop prices, water, electricity and diesel. Decisions are taken far from the village by people who have no sympathy for the farmer or his way of life. Their sympathies lie firmly with corporate India. Should corporate India decide to go into agriculture, it can count on all the support and assistance from these political–economic decision-makers.

Agriculture receives virtually no public investment. The condition of the villages clearly reflects where they stand in the government's scheme of things—pathetic schools, nothing by way of health care, no sanitation, not even water that is clean and safe to drink. Despite occasional noises about rescuing the farmers from the depredations of moneylenders, no law or remedial measure is actually put in place by any party.

EVOLVING PROBLEM

During the mid-1980s, the tragic phenomena of rural suicides was becoming increasingly visible but initially the underlying cause of the suicides was not clear.

From the mid-1980s to the mid-1990s, Punjab experienced a decade of violence at the hands of Sikh militants and the police and army. The majority of rural moneylenders are Hindus and the majority of rural debtors are Sikhs. It is possible that when militancy was at its height, the moneylenders may have hesitated to exert heavy pressure on those who owed them money. Later, when they felt more secure, they may have 'made up for lost time' by exerting even greater pressure to recover their money. This would have driven many debtors to the point of suicide.

It was a common perception that these suicides were being triggered by drug abuse. MASR had tried to control drug addiction in Moonak subdivision with the government's help by arresting drug peddlers. The success of this campaign became visible when, during a visit to a village on the Haryana border, we were summoned by a group of elderly women. The group leader wanted to know what good we had been able to do for the people of the area. While I was still fumbling for words, their angry group leader, a fat woman, intervened to declare that she would tell what good we had been able to achieve for the area. The price of *bhuki-opium-husk* had gone up from ₹2 to ₹4 and she held us responsible for it.

When lowering addiction levels in these villages still did not bring down the level of suicides, we started looking for other causes. It was becoming increasingly evident that the farmers' standard of living was declining. This was affecting the well-being of agro-labourers, as well as all those in the village who were dependent on farm produce.

HOW MASR BECAME INVOLVED

My ancestral village falls in Moonak subdivision. In the course of regular visits to the village, local people initially brought the problem to our attention.

The problem began to surface back in the late 1980s and at first it seemed that the increase of suicides might be just coincidental. But as one factor after another began to weigh in: (*a*) successive years of crop loss on account of bad weather, (*b*) cotton crop loss several years in a row because of American boll weevil infestation, (*c*) rapid increase in cost of seed, fertiliser and pesticide as well as fuel costs, (*d*) sinking water table necessitating re-digging of wells and installation of more powerful (and more costly) pumps, (*e*) dependence on informal and exploitative sources of credit and (*f*) overall rapidly declining profitability. Gradually, it became apparent that farmers were being pushed deeper and deeper into debt and losing their land. The response to the impossibility of ever coming out of this spiral of personal and community distress was suicide. More and more people were killing themselves.

The phenomenon has steadily worsened, with some little respite in the years 2005 and 2006, on account of the rise in land prices.

Documentation Begins

We began to note down the names of suicide victims as well as the dates of suicide in 1988, starting with our own village and gradually including data from more and more villages until we were covering all the villages in the two subdivisions. As people came to know that we were recording this information, they began to contact us on their own with information.

We also started bringing this growing problem to the attention of senior political leaders, persons at the highest levels of government, the universities and the Reserve Bank of India (RBI). K.R. Narayanan, then President of India, was the first to whom we wrote and he responded sympathetically. At his instance, a series of inquiries were initiated. Subsequently we wrote to Dr A.P.J. Abul Kalam, the president who succeeded Narayana. President Kalam also got inquiries made. He praised our research as 'comprehensive'. The RBI conducted its own inquiry which resulted in certain relief measures for those farmers under debt to banks. These measures were conveyed to MASR before the Punjab government was informed and were sparingly advertised—even officers of the Punjab government remained unaware and sometime later, requested for the Reserve Bank's letter from MASR.

The First Shouts for Help

After MASR wrote on 23 May 2002 to President Narayanan, with a copy to the RBI, a response came from the RBI. On 13 June 2002, then General Manager of the bank, Dr Deepali Pant Joshi conveyed the bank's decision to allow farmers to take loans up to ₹50,000. MASR wrote back urging the bank to extend the lending limit to ₹200,000, extend the deadline for application and pointed out that scheme had received very little publicity. RBI was told that some respected agro-scientists had predicted the collapse of agriculture in Punjab by 2020 and MASR agreed with the scientists.

Response to this letter came after a long time. It finally reached on 10 February 2003; the RBI wrote back saying 'We have since issued revised guidelines for compromise settlement of chronic non-performing assets (NPAs) of public sector banks. The guidelines cover NPAs in all sectors irrespective of nature of business, without outstanding balance of ₹10,000 crore, on the cut-off date, that is, March 31, 2000.'

On the face of it, the RBI appeared to be acting very generously and benefiting farm debtors. No doubt, indebted farmers could be helped. But by raising the amount of loan eligible for compromise settlement to ₹10,000 crore it was obvious that the bank actually intended to help the corporate sector. After all, which small or marginal farmers could ever dream of taking a ₹10,000 crore loan from any public sector bank. It would be interesting to find out the amounts calculated in these compromise settlements, in the case of farmers and in the case of corporate debtors.

We have repeatedly written to the elected representatives: all the MLAs and relevant ministers of the Punjab Vidhan Sabha, all the MPs and relevant ministers of the Lok Sabha, and the speakers of the Punjab Vidhan Sabha and the Lok Sabha. Since Haryana is also a largely agrarian state and its farmers face the same problems, we also wrote to the Haryana MLAs and relevant ministers. We went to the press again and again.

Some of those who responded—especially media persons—were skeptical and wanted some proof that there was a 'suicide problem'. To substantiate the cases we were presenting, we asked the village Panchayats to check on the cases and if they found that the person who had committed suicide had been under heavy debt, they recorded their finding on legal stamp paper in the form of signed affidavits.

Figures Tell the Story

During the progress of this study, MASR had documented 1,630 cases of suicide for the period of 1988 to 2008 in the 91 villages of Lehra and Moonak subdivisions. MASR also analysed this data from the following perspectives:

- Date
- Village
- Name

- Occupation
- Age
- Gender
- Mode of suicide
- Landholding
- Debt amount

Over the years, it was seen that the crisis worsened to the extent that in some families, more than one person committed suicide. We have a separate list of these cases along with detailed documentation of the circumstances.

TIMELINE

1988 In 1988, MASR approached the Government of Punjab through the Governor of Punjab S.S. Ray for complete pension coverage for the elderly, widows, disabled and destitute children of Moonak subdivision. This neglected subdivision on the Haryana border was in the grip of agrarian crisis which was manifesting itself through rural suicides. The Governor accepted this proposal and the related departmental heads cooperated. About 9,000 pensions that were granted before the scheme, ground to a halt due to objection by the defeated Congress candidate from this constituency, Mrs Rajinder Kaur Bhattal. She feared the credit would go to the Opposition. However, S.S. Ray—who was a senior leader of Mrs Bhattal's Congress party—reluctantly agreed. These 9,000 pensions helped to restrain suicides for a while. This effort was a precursor to the Pension Plan adopted by Punjab later on.

9 June 1990 On this date, a list of nine suicide cases in the of village Gulahri, dating back to 1988, was sent to A.K. Kundra (IAS, Health Secretary, Government of Punjab). This village is located in Lehragaga block of Sangrur district on the Haryana border. These cases were personally verified by the civil surgeon of Sangrur, who reported that these deaths appeared to have been caused on account of family disputes. In his written statement, he attributed the exact cause of death due to poisoning (pesticide) in some cases and blamed other deaths on heavy consumption of liquor following a domestic quarrel. These are some of the early cases of farmers committing suicides which were confirmed officially. The value of this report was that it confirmed these nine deaths in this village from 1987 to1989 as suicide cases.

1993 Around 1993, MASR was unofficially informed by some friendly senior Punjab government officials that the government was deliberately playing down the suicides, as any talk about increase in rural suicides would give a fillip to the prevailing militancy in the state.

1998 MASR then wrote to the vice-chancellors of Punjab Agricultural University, Panjab University, Punjabi University and Director, Post Graduate Institute of Medical Education and Research (PGIMER), Chandigarh requesting them to conduct research into the phenomena of rural suicide. In the letter, it was stated:

> Since the late 1980s the rate of suicides in Lehragaga and Andana blocks of Sangrur district has been rising. I belong to this area and am in search of ways to reverse this tide of human tragedy. The first step is, of course, to understand why people are taking their own lives. I would like sociologists, psychologists and agricultural economists of Punjab Agricultural University to study this area and its people. As Vice Chancellor, you are in a position to direct the attention of scholars to this backward area of Punjab.

Regrettably, the three universities declined to carry out this research for lack of funds. PGIMER diverted MASR to Rajendra Hospital, Patiala, as the area for health coverage fell under its jurisdiction. The matter ended there.

1998 MASR wrote to President K.R. Narayanan, calling his attention to the growing phenomenon of rural suicides—it listed 93 suicides—and absence of institutional finance. We also endorsed a copy of our letter to the Chief Minister of Punjab, Parkash Singh Badal. President Narayanan's response was immediate and effective. He at once took up the matter with the concerned ministries of the union government, the RBI and the Chief Minister of Punjab.

Letter to then President of India, K.R. Narayanan

April 28, 1998

To the President of India
Sub: Suicides in rural Punjab

Dear Mr. Narayanan,

You are no doubt aware that suicides by farmers in several states have increased to such an extent that it has become a matter of national concern. Reports appearing in the press link these deaths to economic distress arising out of unremunerative prices, crop failure and indebtedness.

I would like to bring to your notice specifically the rising number of suicides in rural Punjab. For one small cluster of five villages in Sangrur district, figures collected since 1994 show 87 farmers and farm labourers have taken their lives. This area of Punjab is typical of the entire state with its 12,000 villages. About 80 per cent of Punjab's population is dependent on agriculture and is predominantly Sikh. If data were to be collected from all of rural Punjab, the suicides figures would certainly be in excess of a minimum figure of 6,000.

From the viewpoint of the individual and his family, suicides are personal tragedies, seen from the viewpoint of society, they are symptoms of systemic problems. Expressed in the simplest terms, we are talking about lack of opportunities and economic injustice.

Rolling back the suicide rate will require many measures but the first requirement is to get a complete and accurate picture of the situation in the entire state. Admittedly, people often wish to conceal the circumstances when a member of the family has committed suicide; those entrusted with the task of enumerating such cases will have to be people who inspire trust and who can deal tactfully with the respondents.

At the same time, attention must be directed to the economic situation which is at the root of the Punjab farmers' bitter desperation. Already a great deal of data is available on crop production, price levels, loans and loan recoveries, industrial growth, literacy and skill generation and all the other markers of economic prosperity or penury. There are several agencies involved in interpreting this data, from the universities to the state Planning Commission. Let these analysts review the statistics with the reminder that poverty can kill.

As President of India, your expression of concern and suggestions to the Union and State governments mean a great deal. We urge you to satisfy yourself regarding the emiseration of Punjab farmers –to the point of suicide and we urge you to recommend to the Union and the State governments a package of measures to correct the economic situation.

Some of these measures are short term, others will have an impact over the next decade.

Enhance the purchase price on the current wheat crop by ₹100, per quintal either directly or in the form of a bonus. This would still be substantially lower than the landed price of imported wheat.

Levy an additional per quintal insurance charge on the government procuring agencies so as to build a fund earmarked for comprehensive crop Insurance. This is a permissible charge in industrial pricing.

Replace the quota system for recruitment to the military and paramilitary services with selection on merit, as was the practice before 1984.

Do away with pegged food prices so that farmers may derive the benefit of liberalisation and sell their grain in a free market.

Make a steep upward revision in land ceilings. Fragmentation has brought landholdings down to an average of 2.5 acres per farm family. It cannot decrease much more. According to farm economists a holding of less than 14 acres is unviable.

The combination of poverty, liquor and drugs is lethal and must be broken. Restructure the state's tax base to enable it to earn state revenues from sources other than excise. At the same time, make a certain level of investment in rural education compulsory so as to raise the market worth of Punjab's human capital. A higher educational level will also pay off in terms of reduction in birth rate.

Designate areas where crops have failed and suicides are taking place as 'distressed regions' and initiate relief measures including pensions to the elderly, destitute, widows, orphans and disabled, and deferment of loan recovery, suspension of interest charges and launch public works schemes to provide employment to farmers and farm labourer facing hard times. Give ₹30,000 as distress relief to the bereaved families.

I enclose a background note and photostat copies of recent newspaper reports of suicides in rural Punjab.

With respectful regards,
Yours Sincerely
Inderjit Singh Jaijee, convenor
Movement Against State Repression
Enclosed list of suicide victims

1998 In early 1998, the then Prime Minister Atal Bihari Vajpayee was on a visit to Amritsar along with Union Minister, Madan Lal Khurana. Chief Minister Parkash Singh Badal, a coalition partner of Vajpayee's Bharatiya Janata Party (BJP), was there to receive him. As reported by the press, Prime Minister Vajpayee enquired from Badal if there were any farmers' suicides in Punjab. Badal denied there were any cases of suicide in the state and added that reports of farmers' suicides were grossly exaggerated.

Later, in the Punjab Vidhan Sabha, however, in response to a question by Hardev Arshi, Communist Party of India (CPI), Punjab Chief Minister, Parkash Singh Badal admitted (*The Tribune*, 3 July 1998) the following suicides:

- 75 suicides (1 January to 16 June 1998)
- 162 suicides (1997)
- 146 suicides (1996)
- 113 suicides (1995)
- 73 suicides (1994)

A Change in Attitude

MASR's reports of 1998 and 2002 brought about a change in the attitude of the media and the government. It was now accepted that there had been an increase in the number of deaths by suicide in rural Punjab. Mention of rural suicides, which was earlier considered a taboo, was now beginning to appear in the press. During this period, district Ganganagar reported 175 rural-suicide deaths in one year. District Amritsar reported 300 deaths during 30 months. A study by the Association of Democratic Rights pointed out 79 suicides in 29 villages in Punjab, out of which 42 suicides were for the period 1997–2000. According to Association for Democratic Rights (AFDR), suicides by Punjab farmers had increased by 250 per cent. Haryana reported a staggering growth in suicide cases, mostly along the state's border with Punjab.

From June to August 2000, 40 suicides were reported from 24 villages of Lehragaga and Andana blocks. Considering that more than half the villages in these blocks had not been investigated for suicide deaths and also that some families preferred to conceal suicides, the total of such deaths for these two blocks works out to around 50 suicides for this period.

Evidence suggested that suicides were more common in Lehragaga, Andana and Barnala blocks. An increasing trend of farmer suicides were noticed immediately after harvesting time. After making appropriate adjustment for these two factors, it appeared very likely that annual suicides in these three blocks were about 80 in each block per year or a total of 240 suicides. Districts Mansa and Bathinda were also badly affected districts. Fourteen blocks of these districts would have 40 suicides each, totaling 560. The annual level of suicide cases in Punjab would be more than 3,000 in the rural sector.

The government's response to these suicides has been disappointing from the beginning. In 1998, the government denied that there were any suicides. In 1999, it admitted suicides but refused to accept that they were largely due to debt and impoverishment. In 2000, it admitted both these factors but failed to provide any direct support to the victims' families. The government policy has been to bury its head in the sand and hope the problem will go away.

According to Dr P.S. Rangi of the Punjab State Farmers Commission,

> What is intriguing are the missing numbers. In 1990–91, there were 2.95 lakh marginal and 2.03 small operational landholdings. In ten years time, by 2000–01, these had come down to 1.23 lakh marginal and 1.73 lakh small operational holdings. A careful perusal would show

that nearly 1.20 lakh farm families had moved out of agriculture in the ten years period. Where have these families gone? What alternative employment opportunities have they adopted? No one knows or even talks about them.[2]

2000 Toward the end of 2000, MASR wrote to the Chief Minister of Punjab and to the Chief Justice of Punjab and Haryana High Court, apprising them of the legal steps initiated under the guidance of Sir Malcom Darling and Sir Chotu Ram by the United Punjab Government in the 1930s, to bail out Punjab's farmers from debt and economic depression.

Letter to the Chief Minister of Punjab, Parkash Singh Badal

November 28, 2000

To S. Parkash Singh Badal
Chief Minister Punjab, Chandigarh

Dear S. Parkash Singh Ji,

During the British Rule, States and Provinces enjoyed much greater autonomy. States had control over their resources and the legislature identified with the population with the result that in case of crisis like the present agrarian crisis, corrective measures were taken promptly.

In the beginning of the 20th century agricultural debt problem had become acute. Moneylenders had started relieving petty farmers of their land in lieu of payment of their loans. In 1929 government appointed a Banking Inquiry Committee. This committee reported that the total volume of loan in the Punjab was 135 crores. It was felt that some relief to the victims of rural debt was urgently needed and a Legislative Council Committee was set up for the purpose in 1932. A bill was later passed called 'The Punjab Relief of Indebtedness Act, 1934.'

---

[2]Dr Rangi, *Report of the Punjab State Farmers Commission*, Government of Punjab, 2006. Also available in R.S. Sidhu, A.S. Bhullar and A.S. Joshi, 'Income, Employment and Productivity Growth in the Farming Sector of Punjab: Some Issues', *Journal of the Indian School of Political Economy* (January–June 2005).

Certain amendments were made to the earlier Acts relevant to the agrarian society. These were the Provincial Insolvency Act, 1920 and Usurious Loans Act, 1918. Another Act called the Punjab Debtors Protection Act, 1936 was later introduced.

Today truncated Punjab's agricultural debt is ₹6,000 crores. The problem is thousand times more serious. Farmers faced with eviction and sell out are going in for suicide in a big way.

In most cases the debt is usurious and illegal but is taken for sheer survival. Such debts need to be settled in a just and equitable manner. This can be done by setting up Debt Conciliation Board, vesting in it the powers that it enjoyed earlier, Through this and other measures suggested in the report, farming can again be made to yield modest profits that are necessary to sustain the 75% rural population of Punjab which today faces abject poverty and dislodgement.

With best regards and Fateh,
Yours Sincerely,
Inderjit Singh Jaijee, Convener,
Movement Against State Repression, Chandigarh
Our report and the summary of the relevant laws are appended.

1. Rural Suicide: A Quantum Jump (Appendix A)
2. Summary of Laws for Regulation of Rural Debt in the 1930s (Appendix B)

This letter to the Chief Minister was followed up by another letter to President Narayanan on 26 May 2002. These letters and reports became the basis of setting up of the Punjab Farmers Commission in 2005 and the Punjab Relief of Agricultural Indebtedness Bill, 2006, brought out by the Punjab government in 2006 (Appendix D).

Three Studies

As a consequence of our letter to the President of India, three studies were undertaken:

1. **Institute for Development and Communication (IDC) study** (1998): The IDC undertook this research on the direction of

the Financial Commissioner, Government of Punjab. In June and July 1998, the IDC conducted a sample survey covering 119 households in seven villages of Punjab's three cultural regions, namely Majha, Malwa and Doaba. Additionally, a qualitative study was carried out; it went into details of 28 cases of suicide from three villages covered by the MASR— Bangan, Balran and Chottian—and from a fourth village, Sakrodi, which is not in Moonak or Lehragaga subdivisions. It may be mentioned here that the IDC report acknowledged the help of MASR and stated that its researchers benefited from MASR's resource material. The IDC researchers fully accepted the validity of the data in the MASR list of suicide cases. The report was published as *Suicides in Rural Punjab*.[3]

2. **Central government enquiry through Ministry of Agriculture**: Vijay Kain in his letter to MASR writes:

> The state government is deeply concerned with the rising indebtedness of farmers in Punjab and on the initiative of the state government, a high level study team constituted by Government of India under the chairmanship of Sh. J.N.L. Shrivastava, IAS, Additional Secretary, Agriculture, visited the state to have first hand information.[4]

Mr Shrivastava's team visited the suicide affected areas of Punjab. Unfortunately, his report was not published.

3. **RBI Inquiry** (1998): The RBI General Manager asked for MASR reports on indebtedness and rural suicides and subsequently, Mrs Teja, a senior RBI official, visited the suicide prone area of Punjab and is said to have substantiated MASR's findings. However, her report was not made public.

Panjab University Study

In the year 2000, Professor Gopal Iyer and Dr Mehar Singh of the Sociology Department, Panjab University, with the help of four research scholars of the department carried out a survey,

---

[3]*Suicides in Rural Punjab* (Chandigarh: Institute for Development and Communication, 1998).
[4]Taken from a letter by Vijay Kain (IAS) Secretary, I.F. Banking (Punjab) to MASR on 19 May 1998.

subsequently published as *Indebtedness, Impoverishment and Suicides in Rural Punjab*. They examined some of the cases reported by MASR and observed that 'intensive verification during the pilot study showed that 67 per cent of the cases covered in the list of MASR were related to debt'.[5]

Verification by Deputy Commissioner (DC) (Sangrur) through Sub Divisional Magistrate (SDM) Moonak

Piqued by the government's obstinacy in accepting the fact of rural suicides, MASR referred 31 suicide cases from 22 villages of Moonak subdivision that happened over a period of six months (April to August 2001) to the Sangrur DC, Sarvjit Singh, IAS. The Deputy Commissioner (DC) of Sangrur instructed the Sub Divisional Magistrate (SDM) of Moonak, Mr R. Tiwari, IAS, to inquire into the cases.

The SDM verified 29 of these cases; He was unable to trace 2 cases as the specific village could not be identified. He confirmed the 29 cases as suicide deaths and further specified the apparent motive behind each suicide. He gave economic distress as the motive for suicide in 26 out of the 29 cases. The remaining three cases were assigned to domestic causes. In 26 cases, the mode of committing suicide was attributed to consuming pesticides, 6 cases were ascribed to hanging. One case was jumping before the train and 1 was due to excessive consumption of alcohol. This is a solid verification of suicide cases by an official of Punjab government and also provides a strong ground for providing financial assistance to the suicide victims.

The english translation of the report by R. Tiwari is reproduced below:

> From: Sub Divisional Magistrate, Munak
> To: Deputy Commissioner, Sangrur
> No. 1084/Steno dated 13/12/02.
> Sub.: Regarding Farmers Suicides in Munak Sub Division.

---

[5]K. Gopal Iyer and Mehar Singh Manick, *Indebtedness, Impoverishment and Suicides in Rural Punjab* (Delhi: Manohar Books, 2000).

Dear Sir,

This letter is in reply to your three letters on the subject,

(1)   Letter No. 253/PA/D.C. dated 28-09-01
(2)   Letter No. 432/Steno/A.D.C. dated 10-01-02
(3)   Letter No. 790/D.A.1 dated 27-11-02

The first letter was about thirty one cases of suicide of which in twenty six cases economic hardship was found to be the cause. You had asked for a detailed and clear information regarding the economic hardship factor. In compliance to your letter the undersigned entrusted the enquiry of these twenty six cases to three Naib Tehsildars. They were instructed to call on each and every grieved family and investigate the cause of economic hardship. These Naib Tehsildars submitted their report to the undersigned on 09-12-02. After studying this report the undersigned has arrived at the conclusion that in the case of these twenty six suicides the cause of economic hardship was due to the undernoted factors:
(1) Some farmers were questioned about the quantum of debt from the Arthias and if they had any record in connection with this debt. The farmers' families declared that there was no written record of the debt and whatever amount the Arthia demanded it used to be paid by the farmer. When Arthias were questioned about this debt they informed that they had full trust in the farmers and similarly farmers had full trust in them and these debts were not committed to writing and the debt was etched in their memory and they knew how much the farmer borrowed and how much he returned. The truth with regard to this is that the Arthias charged the farmers the maximum possible interest rates on the borrowed amount which were well beyond the legal purview. And for the said reason they do not maintain any record of the money advanced and it remains a word of mouth transaction and is carried out through writing on a plain paper (Kachian Parchian). This illegal transaction is also done in order to avoid paying income tax. This is a traditional way of loaning money to the farmers and the farmers are thereby compelled to market their produce through the Arthias. In the absence of any record the farmer has to accept

the word of the Arthia with regard to the quantum of debt. Out of the twenty six cases of economic hardship some cases of suicides are attributable to this reason.

(2) The second reason of suicide is the debt incurred from the banks. This debt is generally incurred to purchase tractor etc., but the farmers' earnings remain insufficient to pay back the instalments. The farmer is then troubled by the fact that the land which he had mortgaged to pay the debt would be taken over by the bank as he was not in a position to return the money. Out of the twenty six cases of economic hardship some cases belong to this category.

(3) In some cases small farmers who have less land, also have less income but their social obligations like marriages, etc., impose an expenditure which they are unable to meet through their own income. The paucity of funds creates friction within the family and this sometime leads to suicide. Out of the twenty six cases of economic hardship some cases are attributable to this reason.

(4) In some families where the parents get addicted to drinking liquor most of their income gets consumed in this indulgence and the children find that they may not be able to get married and in dismay they commit suicide. Some cases out of the twenty six cases of economic hardship are attributable to this reason.

The reports submitted by the three Naib Tehsildars have brought out the above mentioned factors causing economic hardship. I am sending this report along with the report of thirty one cases of suicides out of which twenty six suicides are due to economic hardship. The three investigative reports of the Naib Tehsildars are also enclosed.

Enclosed 29 pages

SDM Moonak

In reply, MASR wrote to the DC of Sangrur:

February 12, 2003, Chandigarh
To: Mr Sarvjit Singh, Deputy Commissioner, District Sangrur
Subject: Relief to next of kin of rural suicide victims

Dear Sir,

Enclosed please find a list of persons belonging to Lehra and Andana blocks who committed suicide out of economic hardship/debt between March 31, 2001 and February 12, 2003.

> We have data on about two-thirds of the villages in these two blocks but we do not claim that this data is complete. It is possible that some suicides are not in our knowledge, likewise some families may have declined to admit suicide as the cause of death out of fear of social stigma or reluctance to get involved in a police inquiry. Suicides subsequent to our visit may have also occurred and have therefore gone unlisted.

While our case for compensation to next of kin is before the Punjab and Haryana High Court, we request you to provide all due relief to the affected families on humanitarian grounds by way of the provisions of pension for widows, elderly and destitute children. In many cases, it is observed that after the death of the bread-earner, the family's economic plight worsens to such an extent that other members of the family see suicide as their only way out. The small sum by way of relief may forestall more suicides in the same family.

Yours sincerely
Inderjit Singh Jaijee, convenor MASR

Eight years later, the response to this letter is still awaited.

State Government Response

2001 In a turn-around in 2001, Punjab accepted that farmers, unable to clear their debts, were committing suicide. The SAD government, headed by Parkash Singh Badal, made a ₹2.5 crore provision in the 2001–02 Budget to provide ₹2.5 lakh compensation per family of suicide victim. It is another matter that, although the government stayed on in power for a year after making this announcement, no steps were initiated to honour this noble commitment. Not a single person was compensated.

2002 Captain Amarinder Singh, leader of the opposition Congress Party, was sympathetic to the farmers when his party was out of power, but when he became the Chief Minister of Punjab in 2002, he ignored the previous government's commitment to compensate the next of kin of suicide victims. The matter was taken to the High Court in 2002. The Congress-led Punjab government informed the High Court that it was setting up the Punjab Farmers Commission to look into the farmers' problems and determine what compensation should be given. This commission, set up in 2005, was headed by an eminent agriculture scientist, Dr G.S. Kalkat. In view of the government's weak financial position, the Farmers' Commission recommended a grant of ₹50,000 initial compensation and a monthly pension of ₹1,500 for 30 years to the next of kin of suicide victims.

Punjab Agricultural University Report

Responding to a request to have the phenomenon of rural suicides in Punjab examined, the then President of India, K.R. Narayanan, referred the issue to the Union Minister of Agriculture who provided funds to PAU to conduct a study. PAU published its findings—*Market Imperfections and Farmers' Distress in Punjab*—in 2003 and made the following acknowledgement in its report:

> Inderjit Singh Jaijee, convener of the Movement Against State Repression (MASR), set off the debate on suicides by farmers in Punjab in 1998. Jaijee wrote in a letter to President K.R. Narayanan stating that 93 poverty-driven suicides, which took place in a cluster of five villages in Sangrur district were the result of a lack of opportunities and economic injustice.[6]

The PAU study dealt with the phenomenon of suicides, identifying causes and suggesting remedies, but did not try to find out how widespread it was or attempt to quantify the suicides.

---

[6]D.K. Grover, Sanjay Kumar and Kamal Vatta, 'Market Imperfections and Farmers' Distress in Punjab', in *Glimpses of Indian Agriculture: Macro and Micro Aspects*, ed. S.M. Jharwal (Delhi: Academic Foundation, 2008).

Punjab's Status Report

In 2004, Chief Minister Captain Amarinder Singh informed the central government in *Punjab Status Report 2004* that '2,116 farmers have committed suicide from 1988 to 2004 but the number could be more'.[7] 'The study by AFDR and others has found that 83.5 per cent were Jat Sikhs, 8 per cent were Scheduled Castes, another 5 per cent were Brahmins and 1 per cent others. Thus, Jat Sikhs plus Scheduled Castes accounted for more than 91 per cent of victims.'[8]

Majority of the Scheduled Castes are also Sikhs.

Criminal Investigation Department (CID) Reports

In 2006, the Punjab government ordered CID to investigate cases of rural suicide in the Lehragaga and Moonak subdivisions of Sangrur district. Although this enquiry was kept under wraps, some information leaked out. According to reliable sources, the CID authenticated 130 cases of suicides from a few villages.

The CID is a wing of the police department and CID investigations are reputed to be thorough—more so than investigations by the regular police. It is difficult to understand why the Punjab government disregarded the CID reports and forwarded only the normal police reports to the Punjab Farmers Commission, which the government itself had set up for the purpose of investigating rural suicides.

Second IDC Study

In 2006, the Punjab Farmers Commission commissioned IDC, Chandigarh, for the second time to conduct another survey on

---

[7]'Suicides by Farmers in Punjab', Unpublished official note of the Punjab State Agriculture Department (Chandigarh, 2004). Also available at http://planningcommission.nic.in/plans/stateplan/sdr_pdf/shdr_pun04.pdf (Online version of *Human Development Report 2004* for Punjab).

[8]K. Gopal Iyer, Mehar Singh Manick, *Suicides in Rural Area of Punjab: A Report* (in Punjabi) (Ludhiana: Association for Democratic Rights, 2000).

Table 8.1:   Farmer's Commission Report (2006)

Year	Estimated suicide rate per lakh population	Punjab rural population 2001	Approximate number of suicide in entire rural Punjab
2001	12.38	160.96 lakh	1,993
2005	13.10	—do—	2,109

Source: Pramod Kumar, 'Suicides in Rural Punjab', in *Punjab State Farmers' Commission* (Chandigarh: Institute for Development and Communication, 2006).

suicides in rural Punjab. This was strange as the second report by the same investigating body was not expected to be materially very much different from the first. This report indicated that the annual rate of suicides in Punjab was around 2,000 a year. An extract from the 2006 IDC report is provided in Table 8.1.

The Farmers' Commission estimated the suicide rate for 2001 and 2005 based on the sample survey of rural population. As per their estimate the suicide rate was 12.38 in 2001 and 13.10 in 2005. The total rural population of Punjab during 2001 was 160.96 lakh. If we project the approximate number of suicides during 2001 and 2005 based on the above Farmers' Commission estimate then the following numbers are obtained.

The Farmers' Commission study also made an attempt to estimate the number of suicides for the years 2001 and 2005. This estimation is based on the sample survey conducted by them and they estimated the annual suicide rate for 2001 as 12.38 and for 2005 as 13.10. It may be pointed out that based on the Punjab police data, they mention that the suicide rate in 2001 was 2.04 and in 2005 it came down to 1.38. When the estimation of the survey data is compared with the police data, the total lack of validity of police data is well established.[9]

Bharti Kisan Union (BKU) (Ekta) Study

At the Bharti Kisan Union (BKU) rally in Bathinda on 7 September 2007, the BKU (Ekta), a farmers' body, released the suicide data (random sample) that it collected using the MASR format (see Table 8.2).

[9]Pramod Kumar, 'Suicides in Rural Punjab', in *Punjab State Farmers' Commission* (Chandigarh: Institute for Development and Communication, 2006).

Table 8.2:   Estimated suicides (2001–05)

	District	Total No. villages	No. of villages sampled	Suicides	Per village average	Extrapolated district total
1	Bathinda	281	70	750	10.71	3,010
2	Faridkot	171	5	25	5	855
3	Ferozepur	1,003	13	77	5.92	5,940
4	Ludhiana	918	5	12	2.4	2,203
5	Mansa	240	46	424	9.21	2,212
6	Moga	329	33	475	14.39	4,735
7	Muktsar	234	16	61	3.81	892
	Extrapolated total for this 8 district sample					29,776

Source: BKU (Ekta) Study.

According to official districts statistics, there are 281 inhabited villages in Bathinda. The BKU (Ekta) team sampled 70, taking all known cases for the period 1990–2006. We see that 750 suicides were documented from 70 villages. This means an average of 10.7 suicide victims per village. Extrapolated for Punjab, the figures come to 90,000.

Punjab Police Report

Toward the end of 2007, a report was prepared by the Punjab Police Department and submitted to the government in 2008. This report claimed that there were only seven cases of farmers' suicides from 2002 to 2006.

The reports evoked skepticism, even within the government. Manish Tiwari reports in the *Hindustan Times*:

Financial Commissioner (Revenue), Punjab, Ms. Romila Dubey, IAS, who is closely monitoring the issue, however, observed: 'There are wide variations in the overall figures received from the deputy commissioners, the state police and some NGOs. The matter would be discussed within the government again before making any recommendation to the Government of India for relief'.[10]

---

[10]Manish Tiwari, *Hindustan Times*, 1 June 2007.

The *Times of India* carried this assessment:

> There is a feeling in the government that more suicides are indicative of the state's inability to handle the crisis, and in turn, a blot on the government's face. But, then, farm crisis is a reality and there are many factors other than just bad state government policies, to be blamed. Rather, it is now time to project the right picture to the centre to get help on the lines of farm packages given to the southern states.[11]

State Revenue Department Report

A report compiled by the State Revenue Department, on the instruction of Punjab's Chief Minister, Parkash Singh Badal in 2008, on the basis of district-wise details received from the deputy commissioners and submitted to the Chief Minister's office (CMO) declared that 132 farmers committed suicide in the state during the past 5 years (2002–06).

Lok Sabha Speaker's Enquiry

On 13 October 2007, MASR wrote to the Speaker of the Lok Sabha, Somnath Chatterjee, enclosing a full report on suicides in Moonak subdivision and Gram Panchayat's signed affadavits pertaining to each case.

The letter to Somnath Chatterjee is as follows:

To: Sri Somnath Chatterjee, Speaker, Lok Sabha   Nov 04, 2007
New Delhi   Chandigarh
Sub: Debt-related rural suicides in Punjab.

Dear Sri Chatterjee,

Farmers and farm labourers of Punjab are under heavy debt, which according to Government figures is more than double the debt in any other state. Crushing debt is driving them to take their own lives. Causes for the past 30 years' downward spiral into debt are many, but a major part of the blame

---

[11]Raminder K. Bhatia, 'Govt, NGO Suicide Figures Don't Match', *Times of India* (Chandigarh edition) 27 June 2007.

falls on an inequitable cost/price structure established by the Centre.

We are listing here seven estimates of the number of suicides in Punjab and one of Munak Sub-Division of District Sangrur. As you will see the numbers range from a mere handful for the past seven years to several thousand. Formulating effective policies requires accurate data. Before a fair and workable scheme can be drawn out it is necessary to know the true level of the problem, in other words, the number of farmers/farm labourers who took their own lives.

Pending a general suicide census for Punjab, we would request you to urge the government to have cases of Munak Sub-Division of District Sangrur examined. These cases have been identified and certified by the Gram Panchayats. A list of 1508 suicide cases with 1408 Gram Panchayat Affidavits are enclosed. This list covers all 91 villages comprising the Munak Subdivision and is for the period 1988 till date. Punjab has some 12,400 villages. We would request you to place the enclosed list and affidavits in the library of the Lok Sabha so that Members of Parliament may verify the data if they so desire.

In 2005 the Punjab government established a Farmers' Commission to look into the situation of the state's farmers, including the prevalence of suicide. This commission made a general study of the problem but did not conduct a census. The Commission, on the basis of the data available to it, concluded that about 2,000 farmers commit suicide each year.

The Punjab State government, on the other hand, collected data on rural suicides through two agencies, the Police and the Revenue Department. It did not involve the Farmers' Commission in this exercise. The Police findings placed suicides in Punjab during the past seven years at only 7 and Revenue Department went a little further and gave a figure of 132 suicides for the same period (as suicide is a penal offence, people are reluctant to record deaths as suicide. This is especially so in Punjab as there has been very little accountability for police excesses.)

Various reports quantifying suicides in Punjab:

1.   MASR projected figure for Punjab 2000:   3,000 per year
2.   Punjab Govt Status Report 2004:   2,116 (period 1988–2004)

3.	Punjab Farmers Commission 2006:	2,000 per year
4.	MASR estimates of Punjab's suicides 2006:	40,000 to 60,000 (1988–2006)
5.	Bhartiya Kissan Union (Ekta) for total Punjab 1990–2006: 90,000 is the extrapolated figure based on 29,766 cases recorded for 8 districts on random check of 261 villages	90,000 (1990–2006)
6.	Punjab Revenue Department's Report 2007:	132 (for past five years)
7.	Punjab Police Report 2007:	7 (for past seven years)
8.	MASR Census of Munak Sub-Division: 2007 (comprising 91 villages and supported by Gram Panchayat Affidavits)	1,508 (1988– to date)[12]

The Central government is willing to concede suicides in the southern states but not in Punjab or Haryana because Punjab has long been projected as an 'agricultural success story'. If it is conceded that Punjab and Haryana farmers are desperate it must mean that agriculture all over India has collapsed.

Punjab is experiencing more suicides than other states because agriculture is overwhelmingly the state's main economic activity. Thanks to the thrust of the Green Revolution it has moved on from one crop to more than two crops a year. Therefore, its losses are proportionally more than other states.

In 2001, We had apprised both the President of India and the Chief Minister of Punjab that suicide is violence turned inward. It will not be long before anger and despair are turned outward and result in social and political turmoil if nothing is done to mitigate the difficulties of the rural sector. The Central government's response has been to multiply the number of paramilitary forces. The increased threat perception to the state is not from so-called 'Naxalites' and 'Maoists'. It is from

---

[12]Since this letter was written, a subsequent survey brought to light an additional 266 cases, raising the figure for the same period to 1,774.

rural emiseration due to unequal distribution of the nation's wealth.

With warm regards,
Yours sincerely,
Inderjit Singh Jaijee, convenor, MASR
Cc: all Members of Parliament

We were not officially informed of the Speaker's response but came to learn that he ordered the cases to be verified. However, from village Panchayats we came to know that government employees had visited the villages and made enquiries about suicide cases and they were carrying a list of cases that matched the list we had sent to the Speaker. Some government employees confidentially confirmed that the inquiry was being undertaken at the behest of the Lok Sabha Speaker. The SDM had received the affidavits of Gram Panchayat sent by MASR as an attachment to its letter to the Speaker.

It is a possibility that what perhaps happened was that the Lok Sabha Speaker forwarded the list—the Gram Panchayat affidavits on suicides that he had received from MASR—to the Punjab government and asked them to respond to the suicide the issue. The Punjab government's response was to check the affidavits' validity through the Revenue Department.

A few revenue officials approached by MASR workers stated that more than 90 per cent of the cases were found to be correct. When the affidavits were found to be correct, the state government had to find a way to evade embarrassment. We came to know that senior officers in the Revenue Department returned the verified lists to the Subdivisional Revenue officials with instructions to ignore all cases that were not mentioned in police records or where the families could not produce medical or death certificates.

Aside from a desire to avoid embarrassment, the state government was also at pains to bring the number of accepted cases down to a minimum because MASR had petitioned the Punjab and Haryana High Court for compensation to the suicide victims' next of kin, and the court was soon to rule on the matter.

There was hardly any case that met the police record and death certificate requirements and nothing more was heard about the Speaker's enquiry.

Punjab Agricultural University (PAU) Report

In May 2008, the SAD government of Punjab directed PAU to conduct a survey of rural suicides in the state from 2005 to 2007. They were to enumerate the suicides of farmers only.

PAU invited academics and experts for a conference at Ludhiana to know their views on the conduct of the census. It was the near-unanimous opinion of the experts that the period of appraisal should be extended from 3 to 10 years and it should cover the farming community as defined by the National Farmers' Commission. The survey was to begin from 5 September. Apparently enlarging the period of survey to 10 years and adopting a definition of a farmer covering a larger section of the people has not gone well with the government. The PAU researchers originally intended to take the census for a 10-year period and define 'farmer' in such a way that it would cover a large section of the rural community. The state government held up the financial allocation for the research until the criteria were redefined so as to bring a smaller population under the study.

In August 2008, a team under Professor Ajit Singh conducted a pilot survey covering a few villages of districts Bathinda and Sangrur in order to assess cost and men requirement for the census. The government sanctioned a small amount for the pilot project. Thereafter work on this census survey was stalled.

In early November 2008, PAU announced that it was about to start the census survey of farmers' suicides in the two districts of Sangrur and Bathinda. The survey would cover both farmers and farm labourers for the period 2002–07. Informally we were advised that this census would take about 4 to 5 months to complete. We once again sent the lists of suicides of Lehragaga and Moonak subdivisions to PAU and to the Deputy Commissioner of Sangrur. Copies of the lists were also sent to the Gram Panchayats with a request to present them to the enumerators when they visited the villages. These lists

Table 8.3: BKU (Ekta) suicide survey results (2007)

	Farmers	Agri-labour	Total
Bathinda	773	483	1,256
Sangrur	984	659	1,643
	1,757	1,133	2,890
	Due to debt	Other reasons	Total
Bathinda	827	429	1,256
Sangrur	932	502	1,434
	1,959	931	2,890

Source: BKU (Ekta) Suicide Survey, 2007.

pertained to 1,614 cases out of which 1,542 carried Gram Panchayat affidavits.[13]

PAU Findings

PAU finally started the census at the close of 2008 and extended the date of suicides to be tabulated from 2005–07 to 2000–08. A sum of ₹9 lakh was allocated for this census which was to be carried out in the two districts of Sangrur and Bhatinda. This information was to be verified from *sarpanch*s (head of the Panchayat), other elected Panchayat members or one or more elders of the village for correctness. The findings are listed in Table 8.3.

The PAU report observed that 'The problem is very acute in the Lehra and Andana blocks', comprising the Lehra and Munak subdivision of district Sangrur.

Sample Studies and Census-based Studies

Except for the studies by MASR, the BKU study of suicides in seven districts of southeast Punjab in 2006 and the PAU census of the two

---

[13]MASR continues to record cases of debt-related suicides in the 91 villages of Moonak and Lehra subdivision. Up to November 2010, we have verified reports of an additional 232 cases which occurred in the period between 2000 and 2008. These cases have been included in our total data which forms the basis of analysis in terms of year, age, occupation, gender, holding and quantum of debt (For more information, see Chapter 9).

districts of Sangrur and Bhatinda conducted in 2009, all the other studies are based on the sample method.

Panjab University and research institutes are very poorly funded for undertaking research work, with the result that they tend to do their research work by taking small samples. This is clearly exemplified in the case of the two studies on the same subject conducted by IDC. There is wide variance in the observations of suicides between their 1998 and 2006 reports. In the 1998 report, district Sangrur is shown as the most suicide-prone district in Punjab, with district Faridkot rated at the fifth position. The position is reversed in the IDC 2006 report. District Sangrur is relegated to the seventh position and district Faridkot upgraded to the first position. Interestingly, Faridkot happens to be the district of Punjab chief minister Parkash Singh Badal; his ancestral village is in that district and he has been elected from Lambi assembly constituency of Faridkot district.

Beyond Numbers

National Crime Records Bureau statistics for all of India say that close to 2,00,000 farmers have committed suicide in India since 1997. However, no officially accepted figures were available for Punjab until PAU's study of the suicides from the two districts of the state was published. Although Punjab is responsible for producing nearly two-thirds of India's grain, the desperate plight of its farmers since the mid-1990s had been studiously ignored.

The high suicide rate in Moonak subdivision is one of the many symptoms of a deep and pervasive malaise. The wider picture must take the following into account:

- Economic loss: Increased pauperisation and reduction in productivity.
- Disappearances: Many men, who are deeply in debt, flee the village without telling anyone. In many cases they are never seen again. Their families sink deeper into economic distress and suffer social stigma.
- Monetisation/commodification: This practice is obliterating traditional ethics and practices. The bright side, however, is that monetisation is weakening the caste system. The dark side

is that bonds within the village community, which were a kind of insurance against hard times, are increasingly forgotten.

Pauperisation leads to the following defects:

1. Dependents such as aged parents and children (particularly girls) are abandoned.
2. Mothers occasionally desert their families after husband's suicide.
3. Widowhood is a great economic and social disability in Punjab villages. Ideally a widow should have a place in the deceased husband's family. This rarely works out. Her own family also would rather she was somewhere else. Widows typically take up agricultural labour or domestic work in the homes of better-off families. Exploitation of widows is becoming common.
4. Pauperisation gives impetus to the practice of purchasing women from other, poorer states.
5. It gives impetus to female foeticide.
6. Traditional inheritance laws and practice of giving dowry work against girls.
7. Change in the demographic/cultural character of the village.

VILLAGE PROFILES

Sangrur district ranks third in the state after Ferozepur and Faridkot districts. District Sangrur comprises 697 villages out of which 693 are inhabited. Additionally, it has 17 towns with municipal corporations. Government canals cover 1,41,000 hectares; additional irrigation is provided by 2,99,000 private tube-wells. The total cropped area of the district is 8,63,000 hectares, with 678 villages where water scarcity issues have been identified.

Demography

District Sangrur till lately was divided into six *tehsils*: Sangrur, Sunam, Malerkotla, Barnala, Dhuri and Moonak. The *tehsil*-wise

Table 8.4:   Punjab findings

Name of the tehsil	Area (in hectares)
Sangrur	78,079
Malerkotla	69,536
Bamala	1,40,965
Sunam	93,617
Moonak	60,286
Dhuri	59,934
Sangrur district total	5,02,417

Source: *Farmers' and Agricultural Labourers' Suicides Due to Indebtedness in the Punjab State: Pilot Survey in Bathinda and Sangrur Districts,* Department of Economics and Sociology, Punjab Agricultural University, Ludhiana, 2009.

area of the district, according to the Deputy Economic and Statistical Advisor, Sangrur, is given in Table 8.4. Sangrur district has 5,02,417 hectares.

The rural population of Moonak subdivision constitutes 75.1 per cent and 10.62 per cent of the rural population of the district. Figures are given in Tables 8.5 and 8.6.

For the purpose of better comparison between the villages in Moonak *tehsil,* the villages have been divided on the basis of blocks, that is the Lehra and the Andana blocks.

Geographical Factors

Some villages have fewer suicides than others and there are two reasons for this. One reason for this is access to irrigation. Lehra

Table 8.5:   Moonak subdivision

Male	Female	Total
50,333	69,984	1,20,317

Source: *Statistical Abstract of Punjab,* 2003.

Table 8.6:   Sangrur population as on 1 January 2004

Rural	Urban	Total
14,15,354	5,84,819	20,00,173

Source: *Statistical Abstract of Punjab,* 2003.

and Moonak subdivisions are divided into three distinct areas, namely the trans-Ghaggar areas along the Haryana border, the area in the flood plain of the Ghaggar and the area at the very end of the Ladbanjara and Boha Canal distributaries. The greatest numbers of suicides are found in the canal 'tail end' villages, followed by the trans-Ghaggar area.

Ghaggar flood-prone villages have the fewest suicides. This is because the residents of flood-prone villages are accustomed to getting only one crop per year on account of floods and have adjusted their lifestyle to this single crop economic status. Double-cropping became possible only recently and even now a second crop is regarded as a bonus. Because of the high water table in this area, the cost of production is less than in the other villages.

Cultural Factors

Analysis of suicide data on village-wise basis revealed an unexpected aspect, namely that those villages inhabited by large numbers of originally West Punjab families, that had been resettled there after Partition in 1947, had fewer suicides than the villages where families had been settled in the same place since time immemorial. One way to explain this would be to say that the hardships that the uprooted families endured in 1947 and the years thereafter, acted as a kind of 'natural selection'. The weak perished, leaving only those who were strong and possessed unfaltering work culture. This hardy spirit continues in the second and third generations.

Although all the villages in these two subdivisions are within the borders of Punjab, some of them are inhabited by people whose ethos and traditions are more typical of the adjoining state of Haryana than that of Punjab.

Looking at the village community as a whole, the overwhelming majority of those who committed suicide have been Sikhs. Among the categories of Other Backward Class (OBC) and SC, one finds many who are Sikhs.

Traditional artisan communities such as the Lohars (blacksmiths), whose services were once vital to the village, have suffered greatly over the past few decades as modern manufacturing has marginalised them. Many Lohars are Muslims and this too has contributed to their depressed status. They are the poorest of the poor and suicides among them are vastly disproportionate to their numbers in the village community.

To get a better idea of the nature of the villages, the help of Census of Punjab 2001 has been referred to get various other important indicators of the villages. At the time MASR wrote to the Lok Sabha Speaker, the 1988–2007 tallies came to 1508. By mid-2008, the total numbers of cases recorded from the 91 villages of *tehsil* Moonak (comprising Lehra and Andana blocks) had climbed to 1,583 with 1,566 Panchayat Affidavits. The youngest victim was 18 years old while the oldest was 70 years plus.

PAU STUDY: A CRITIQUE

The state government was unhappy with the census as numbers reported were very close to what the NGOs and farm unions had been citing, and far above the numbers cited in reports by the state police and deputy commissioners. After the census came out, the state government flatly announced that, regardless of the evidence, it would not allow the state to be declared 'suicide affected'. The logic for this refusal was not explained.

The NGOs and farm unions were both happy and unhappy with the census, which incidentally was the first officially commissioned study to employ the form of a census.

On the positive side, the NGOs and unions were happy that the PAU researchers arrived at nearly the same number of suicides for the same area and period (2000–08) as they had documented. If researchers employed in a government-funded university were going to falsify results for anyone, the falsification would have served the government's interest. The PAU figures, arrived at independently, made the case for remedial action very strong and it became impossible for critics to dismiss the problem as imaginary or reflecting bias or ulterior motive.

The census of rural suicides conducted by the PAU, in early 2009, provided a fairly reliable picture of the suicide phenomenon in Sangrur and Bathinda districts. The PAU survey blamed 86 per cent of rural suicides on economic hardship and debt. Earlier, in 2004 the Punjab government submitted a status report on rural suicides to the

Table 8.7:   MASR's suicide figures compared with PAU study (2009)

MASR's number of suicides (2000–08)	969
MASR's Gram Panchayat affidavits	906
PAU's number of suicide (2000–08)	546
MASR and PAU common cases	428
PAU additional cases	118
Number of villages/*mandis* not covered by PAU	9/3
Cases covered by MASR in above 9 villages/3 *mandis*	143
Total PAU records matching with CID reports	66
Total PAU records not matching with CID reports	59
Total PAU records matching revenue report	22
Total PAU records not matching revenue report	6

Source: Author.

Centre, mentioning only 2,116 suicide deaths since 1988. This was a gross under-reporting. The PAU study is certainly better.

Nevertheless, there are certain methodological shortcomings in the PAU study that need to be improved upon before conducting rural suicide census in the remaining districts of the state. These can be seen by comparing the PAU census with the documentation done by MASR over the past more than 20 years.

At the time the PAU census was underway, MASR had identified 969 cases of suicides in 91 villages of Moonak division in Sangrur district (period 2000 to 2008 only). The PAU study has considered extension of two large villages as two separate villages and therefore there number of villages is 93. The PAU study identified only 546 cases; apparently 423 cases less than MASR. The period under study is 2000–08. Exact figures are given in Table 8.7.

Additional Cases Identified in PAU Study

Out of 546 cases identified in the PAU study, there are 125 additional cases of suicides specified by them that are not included in the MASR figures. This 125 constitutes 22.89 per cent of additional cases. If we add 22.89 per cent (say 23 per cent) of these cases to the MASR figure, the total number of suicides will go up to (969 + 223 [23 per cent addition]) = 1,192.

Why PAU's Data Falls Short of MASR's Data

The methodological approach to the study of identifying suicides by PAU mentions that the data was verified either from *sarpanch* or other elected members of the village Panchayat or one or more elders of the village.[14] The problem lies in the fact that the *sarpanch*, all the elected members of the village Panchayat and the elders of the village did not sit together at one point of time to collectively arrive at a decision of validating individual cases, as well as all the cases of suicides which have taken place during the entire period of 2000–08 in the village. The presence of investigators of PAU during such focused group meeting affords a reliable basis of obtaining the accurate number of suicides. This method was not used by PAU investigators. This is a foolproof system of obtaining accurate number of cases of suicides. Further, it also affords the opportunity to the investigator to probe into the veracity of the actual number of cases of suicides ward-wise, which is the smallest unit of village Panchayat. It could simultaneously be verified with other members who are sitting in the focused group situation.

However, the PAU study must be commended. Given the constraints of time, manpower and budget at its disposal, PAU went deeper than all previous studies. Their data is in marked contrast to the figures submitted by Punjab Police (7 rural suicide deaths for the entire state of Punjab) and by the revenue officials (132 rural suicide deaths for the entire state of Punjab) for almost the same period.

Other Reasons for the Difference in PAU and MASR Data

There are also other reasons to substantiate why PAU's data could not capture reality. These reasons contributed to the accentuation of differences in PAU and MASR data.

1. There are at least 56 cases which have been documented and verified as suicide cases out of the MASR's list by CID and these cases are also covered by PAU in their study. However,

---

[14]*Farmers' and Agricultural Labourers' Suicides Due to Indebtedness in the Punjab State: Pilot Survey in Bathinda and Sangrur Districts*, Department of Economics and Sociology, Punjab Agricultural University, Ludhiana, 2009, p. 3.

the PAU data does not include 47 cases identified by CID as farmers' suicides but all these cases are also included under MASR figure. The exclusion of 47 cases by PAU again indicates the limitations in PAU study.

2. The revenue department has authenticated 28 cases (out of 31 cases given to them by MASR). The PAU data identifies 18 of them as suicides and does not include the remaining 10 cases which have been identified as suicide in Revenue Verification Record (out of MASR's data as suicide cases). These 10 cases have been missed out as suicide cases in PAU study.

3. PAU study has not covered six villages (Fathegarh, Gobindgarh Jejian, Kadail, Khanauri Kalan, Sangtiwala and Phuleda) and three rural *mandis* of Lehra, Moonak and Khanauri; these *mandis* have large chunks of agricultural land attached to them. The MASR data shows 143 cases of suicides in these six villages and three *mandis*. This is a big omission in PAU's data. PAU's list includes 125 cases that had not come to the attention of MASR. This indicates that thorough door-to-door census is required.[15]

The above three points substantiate the reason for the differences between the MASR and PAU data. MASR had prepared village-wise suicide lists through a process of joint consultation and focused-group interview with the *sarpanch*, Panchayat members and elders of the village. These village-wise suicide lists—authenticated by the village Gram Panchayats—were provided to PAU. PAU had pointed out 118 cases missed out by MASR from these 93 villages. A research scholar of Harvard University (USA), Malika Kaur investigated suicides in MASR and PAU-examined village Chottian and added three more suicides to the MASR list. This goes to indicate that a more focused investigation needs to be undertaken.

Limitation in the Methodology of PAU

It appears that the investigators of PAU tried to collect the data on suicides, as indicated, from a single source of information. A single

---

[15]It is possible that PAU did not find any suicide cases in these six villages and three *mandis*.

person may not be knowledgeable about all the suicides in the village for nine years.

It also appears that the PAU investigators visited each village only once. Suicide is a sensitive subject; villagers will not like to open up until rapport and empathy have been established with the investigators. Reliable data is unlikely to be revealed in one visit. It is an accepted methodological procedure in social sciences that sensitive and in-depth information not only needs several visits to the village, but also requires adopting focused-interview method to know the overall situation in depth and in totality. PAU investigators apparently have not followed focused-group interview method.

The village society in Punjab is divided on the basis of caste and clan (*pattis*). In the villages under study, very few persons other than Jats or Scheduled Castes have committed suicide. If the *sarpanch* is a Jat he may give information pertaining to suicides by Jat villagers, as he would be intimately acquainted with them. If the *sarpanch* is a SC, he may convey information more about SC villagers. This slant and bias in information may or may not be deliberate. Similarly, clan bias can also creep in. This is one of the shortcomings involved in approaching a single individual for overall information in the village society. In this scientific methodological study, it is very essential that the investigator should clearly and explicitly establish reliability of the data he collects. This has not been done in the PAU study.

The PAU study is the first census study of farmer's suicides in Punjab. It is a rushed census as it has not covered all the households in the village in the course of the survey. This is apparent from their report where they do not specify the number of households covered from each village and whether all households of the village were covered.

Non-debt cases indicated by PAU are unrealistic. The debt incurred in the case of institutional borrowing is generally known to everybody. The main source of debt is from non-institutional sources primarily like the *arthiyas*, but also from friends and family members. These are confidential sources. The *arthiyas* press their dues as and when favourable opportunity arises. At times even the family members learn about this debt much later. The PAU report has indicated that 'there is no definitive method to quantify the level of debt leading to suicide'.[16] The PAU assessment has been harsher in the case of labour suicides.

The major criticism involved the distinction made between 'debt' and 'non-debt' suicide. Cases in which the victim or his family retained at least some land, which they could have sold to clear their debts, was regarded as 'non-debt' suicide. There was no justification for this view. It is true that people often committed suicide because they had come under many pressures, of which debt was one. These other pressures did not mean that the pressure of debt was less or inconsequential in their decision to take their own life. The NGOs and farm unions urged that the next of kin of all 2,890 suicide victims—PAU total figures for districts Sangrur and Bathinda—be deemed eligible for compensation and not just those cases deemed to be debt-related. The PAU findings showed that in the case of 'other reasons' also, the average indebtedness was substantial.

The cut-off date was poorly chosen. The PAU study covered the period 2000–08. The Farmers' Commission Report (2006) points out that annual number of suicides began to rise abnormally from 1992. From 1997 onwards, the number was extremely alarming (see Table 8.1).

Cases from the entire decade of the 1990s should be eligible for compensation. Some other studies used 1997 as the cut-off date. In no case should the cut-off date be later than 1997.

PAU has taken the cut-off point of farmers' suicides from 2000. MASR has recorded such cases of suicides systematically village-wise for Moonak and Lehra subdivision for each year from 1988–99 and 2000–08.

As per the MASR data (updated to November 2010), the number of suicides during these two periods works out to 1,685 as shown below in Table 8.8.

If 2000 is taken as the cut-off point instead of 1991, then injustice is meted out to 614 families who are victims of farmers' suicides which have taken place between 1991 to 1999. In addition, as per MASR's records, another 89 persons committed suicide in the period between 1988 and 1990. Taking 2000 as cut-off is an affront to social

[16]*Farmers' and Agricultural Labourers' Suicides Due to Indebtedness in the Punjab State: Pilot Survey in Bathinda and Sangrur Districts*, Department of Economics and Sociology, Punjab Agricultural University, Ludhiana, 2009, p. 3.

Table 8.8: MASR suicide data

Period	Number of cases
1991–99	589
2000–08	1,096
Total	**1,685**

Source: Author.

justice in respect of the total affected families. Hence, MASR argues that the cut-off period for the census of farmer suicides for the purpose of providing relief should predate 2000 by at least a decade.

This cut-off point should be adhered to also in respect of districts other than Bathinda and Sangrur, for which the Punjab government is launching a census studies. It needs to be extended as a real census study, covering all the households in the village.

The state government intends to adopt a census method similar to the PAU census in the remaining districts for identifying rural suicide cases. If the shortcomings of the census are not rectified, then the data from other districts will be liable to the same criticisms. MASR urges the state government and the experts to review the methodology and make corrections before launching census in other districts.

For steps to make Punjab's rural suicide census more reliable, cost-effective and quicker, the following methodology (Table 8.9) is recommended for adoption:

Benefits of This Methodology

1. The work of door-to-door census has been assigned to the respective Gram Panchayat members; they represent the respective wards of the village.
2. The entire exercise of all Panchayat suicides census can be completed within a period of 4 to 6 months.
3. This method would be cost-effective as well as accurate. Stationary and postal charges would be the main expense.
4. The Punjab census of rural suicides would serve as a model for census in other states. An all-India suicide census can be completed in about 6 months.

Table 8.9:   Methodology: Seven points

1	The state government to prepare a census format and forward it to the districts	4 weeks
2	Deputy Commissioner to forward the census format to respective subdivisional magistrates to be distributed to the Gram Panchayat	2 weeks
3	Gram Panchayat to record suicide data	2 weeks
4	The list of suicide cases so collected to be signed by all the members of Gram Panchayat and displayed on the village paths to invite objections or deletions	3 weeks
5	List to returned to the Deputy Commissioner	1 week
6	Deputy Commissioner to enter suicide data in the database and forward list to the designated state level authority	2 weeks
7	The state government to complete the database for all districts	4 weeks
	Total	18 weeks

Source: Author.

5.   This procedure would make it easier to ensure justice to the suicide-affected families.

In its reply to the Punjab and Haryana High Court in the case of MASR (I.S. Jaijee vs the State of Punjab, Relief to Farmers and Farm Labourers), the Punjab government has expressed an aversion to claiming 'suicide-affected' status for the state. No grounds for this policy have been given. This 'aversion' amounts to deliberate concealment of rural suicides. And yet the government's own commissioned studies show that an abnormally high rate of rural suicides is seen in Punjab.

The Government of India has accepted the states of Andhra Pradesh, Karnataka and Maharashtra as suicide-affected states for the purpose of central government relief but it does not include Punjab in spite of the fact that the number of suicide cases in Punjab is much higher than in those states. The Punjab Farmers Commission in its report has strongly recommended that Punjab be accepted as a suicide affected state like Andhra Pradesh, Karnataka and Maharashtra.

The Punjab government's deliberate refusal to declare Punjab a suicide-affected state amounts to violation of the citizen's Right to Life, enshrined in the Indian Constitution. The Punjab government should be honest enough to accept its responsibility for farmers placed in such a gruesome plight and should come forward like states of Andhra Pradesh, Karnataka and Maharashtra, which have accepted the suicide-affected status. The state of Punjab should face facts and get help from the Centre, commensurate with what has been given to other states. Not to do so is an injustice to the farmers of Punjab.

The incidence of farmers' suicides is much higher in Punjab compared to the other five states—Maharashtra, Andhra Pradesh, Karnataka, Madhya Pradesh and Chhattisgarh—that have seen a high number of farmer suicides.). This can be seen in Table 8.10.

Table 8.10: **Punjab and five suicide-ridden states**

State	Farmer suicides 2000–08	Rural population (cr)	Farmer suicides per lakh population
Maharashtra	34,659	5.58	62
Andhra Pradesh	18,396	5.54	33
Karnataka	20,592	3.49	59
Madhya Pradesh and Chhattisgarh	25,137	6.10	41
Punjab	24,732	1.61	154

Sources: (i) Farm suicide figures for Maharashtra, Andhra Pradesh, Karnataka, Madhya Pradesh and Chhattisgarh are from P. Sainath. 2010. 'Farm Suicides: A 12-year Saga', *The Hindu*, 25 January.

(ii) Punjab figures are from PAU study entitled *Farmers and Agricultural Labourers' Suicides Due to Indebtedness in the Punjab State: Pilot Survey in Bathinda and Sangrur Districts*, Department of Economics and Sociology, Punjab Agricultural University, Ludhiana, 2009.

9

A Problem Does Not Exist if You Don't Look at It

In 2008, Sharad Pawar, Union Minister of Agriculture was asked why Punjab was left out of the ₹4,000 crore package given to the southern states for rehabilitation of families of suicides victims. He answered that the Punjab Chief Minister, Parkash Singh Badal had informed him that only three farmers' suicides had taken place in the state. This was strange because Pawar himself had personally given a ₹1,00,000 compensation to each of the five families whose breadwinners had committed suicide in Punjab. From this one can only infer that while the Union Minister is prepared to grant relief to the next of kin of rural suicide victims in south Indian states, he does not wish people in identically miserable condition in Punjab to receive any share from this central assistance. The central government does not wish to concede that the most agriculturally progressive state in the country has sunk to a point where it is experiencing rural suicides on a mass scale.

Cases of rural suicides have being brought to the government's notice from the mid-1980s onwards. The President of India, in response to the reports sent to him by Movement Against State Repression (MASR), brought it to the notice of the Government of Punjab and the Government of India at least five times. The Union Ministry of Finance, the Union Ministry of Agriculture and the Planning Commission were being notified by the President of India from 1998 onwards. In 2007, the President again thanked MASR for its 'comprehensive report'.

WHERE THE TRUTH WILL HURT AND WHY

It is our impression that the government does not wish to acknowledge suicides and instead of carrying out an honest assessment, it has been handing out research assignments to small research units like the Institute for Development and Communication (IDC) or universities that lack the resources to carry out a state-level census. These studies are directed more towards finding the reasons for suicides rather than finding the level of suicides and to understand the seriousness of the problem. These institutes are largely dependent on government assignments and therefore are careful not to rubbish the government position over-much. Because of this, on 5 September 1998, MASR wrote to S. Parkash Singh Badal, asking him to entrust the work of investigating rural suicide to the Punjab Agricultural University (PAU) or a central organisation, such as the Central Bureau of Investigation (CBI).

If the work of investigation is not given to small research institution or university departments, the government directs the revenue or police departments to tabulate the suicide figures recorded in their files. Three deputy commissioners of Sangrur, Bathinda and Mansa, respectively, as well as the Sub-divisional Magistrate (SDM) of Moonak, on a single day declared to the press that there were no cases of suicides in their respective districts. There was obviously pressure on the administration to downplay suicides.

There are several reasons why the state government, and particularly the police, are reluctant to come out with an honest enumeration.

During the period of Punjab militancy (1984–94), there were a large number of unrecorded killings and reports of disappearances and cremation of unidentified bodies at the various crematoriums of the state. Around Amritsar, in three crematoriums, 2,100 bodies were cremated as unidentified. Later the CBI, on the orders of the Supreme Court, carried out an investigation and identified approximately 1,400 out of 2,100. In these cases, recording cause of death in the death and birth registers would place the police force in a difficult situation. This is why post mortem of unnatural deaths was not made mandatory unless there was a police case.

Another reason why suicides go unreported is that suicide is a penal offence. Families are afraid to disclose suicide deaths lest they give the police an opportunity to harass them.

Social stigma is attached to suicide. Families conceal such cases to protect their standing in the village.

In the majority of cases, the immediate cause of death is poisoning and the most easily available poison is pesticide. Some people believe that farmers do not deliberately drink pesticide or consume Celphos tablets. It is argued that these substances are consumed accidentally or inadvertently in the course of agricultural operations. Dr Inderjit Singh Dewan of the Post Graduate Institute of Medical Education and Research (PGIMER), Chandigarh, has studied this question and concluded that in his 27 years as a doctor he had seen only four genuine cases of death caused by accidental inhalation of pesticide vapour and all the other scores of pesticide deaths he had seen were suicides.

These views were also reflected in a press report. Sarabjit Pandher,[1] a correspondent for *The Hindu* described the reluctance of the next of kin of suicide victims to acknowledge suicide as the cause of death. Pointing out that suicide is an offence under the Indian Penal Code and requires police to register a report, Pandher wrote that fear of the police compels people to conceal the actual circumstances of death. If the police become involved they may threaten to implicate the family in a case of abetment.

Another reason is that they fear the police will take the body away and not return it. The family members fear that the body will be subjected to an autopsy in which organs are removed and the body is mutilated. In such a situation, they are afraid of not being able to perform the last rites of the deceased. Also when someone commits suicide, the family is stigmatised and it may become difficult to get daughters married.

Pandher visited several villages in the southern region of district Sangrur where suicides are endemic. He wrote that not one of the more than 80 cases of alleged suicides during the last 5 years in five villages was reported to the police. The elders in the villages said it

[1]For more details, see Sarabjit Pandher, 'Suicides Are Reported as Accidents', *The Hindu*. 22 April 1998.

was preferred to dispose of the bodies before the police come to know about the incident. Later the cause of death could be given as an 'accident', fever or even snakebite.

Yet another reason Pandher discovered was economic. The Punjab Marketing Board (known as the Mandi Board) pays ₹20,000[2] to the next of kin of those who die while performing farming operations. By claiming that a person died on account of an accident, the family can hope to claim the compensation. Pandher cited the example of a man who had actually ended his own life by drinking pesticide. His family insisted that he had inhaled the toxic chemical while spraying the fields and successfully claimed the compensation.

However, an officer at the Primary Health Centre at Moonak told Pandher that during his 5 years of service in the area he had not come across a single case of death resulting from inhalation of pesticide while spraying. The view of this rural doctor was supported by Dr Inderjit Dewan who is also a renowned forensic expert from the PGIMER. Dr Dewan pointed out that the immediate symptom of toxic inhalation is nausea and most victims can be saved by timely treatment in a hospital. This is not so in the case of consuming Celphose—a cheap and easily available highly toxic fumigant used in the storage of wheat. According to Dr Dewan, ingestion of Celphose kills within 30 minutes of its intake.

Consuming Celphose is the most common mode of committing suicide in Punjab.

THREE STATEMENTS

1. The Division Bench of the Supreme Court, comprising Justices Kuldip Singh and Saghir Ahmed, pronouncing their order, directed the CBI to continue probing charges that the Punjab Police surreptitiously disposed of thousands of bodies in the pre-1995 period.
2. The memorandum submitted by the Punjab State Magistracy to the Governor of Punjab, dated 28 August 1998 says: 'If

---

[2]This amount has since been increased to ₹2,00,000.

we go by the newspaper reports that such and such terrorist were involved in so many hundred killings and if we add up the figures of the past 2 years then the number of innocent persons killed would run into lakhs'.[3]

3. Chief Minister Beant Singh admitted in the House (Punjab Vidhan Sabha) that from January 1991 to the end of 1992, 41,684 Punjab policemen were given monetary awards for eliminating terrorists.

WHY THE GOVERNMENT (AND PARTICULARLY THE POLICE) WANT TO AVOID AN HONEST SURVEY

How the police and politicians shirk their responsibility to investigate deaths in Punjab is exemplified by the various figures on deaths given at various times. In an article published in *The Tribune* on 10 July 2001, former Punjab Director General of Police (DGP) P.C. Dogra refers to '12,000 precious civilian lives' lost during the period of militancy. His predecessor, former DGP K.P.S. Gill, however, sets the figure sometimes as 'low' as 15,000 civilian deaths and sometimes hiking the figure to 35,000 civilian deaths. During the years of Congress rule in Punjab, the party's election manifesto cited 30,000 civilians killed. A senior Supreme Court lawyer, Ramaswami, who was representing Punjab police officers accused of illegal deaths, submitted to the CBI court, a figure of 55,000 civilian deaths during this same period. Neither of these estimates takes into account the number of disappearances from the countryside or the missing prisoners or the security personnel. Punjab's civil magistracy in its memorandum to the Governor of Punjab, citing police reported claims, set the figure at more than 2,00,000. Estimates made by human rights organisations of Punjab concur with the figures put forward by the state's magistracy. However, whatever the correct figure may be, neither the state nor the central government is interested in investigating.

---

[3]Memorandum submitted to Sri Surendra Nath (Governor of Punjab) by the Association of Punjab Civil Services, Chandigarh, 1993. Also reported in 'Officers Demand Enquiry', *The Tribune*, 28 August 1993.

Investigating and intelligence agencies are available to the Punjab government, including the CBI, the police, the Criminal Investigation Department (CID), the Intelligence Bureau (IB), etc. The central government has a number of intelligence agencies operating in Punjab as Punjab is a border state and is a sensitive state. Why are both the state and the central governments feigning ignorance of rural suicides in the state?

Previous presidents of India, K.R. Narayanan and A.P.J. Abdul Kalam have forwarded MASR reports on Punjab's rural suicides to the Prime Minister, the ministries of finance, agriculture, the Planning Commission and to the CMs of Punjab. Both presidents have described the reports as 'comprehensive'. Neither the union government nor the state government can truthfully claim to have been ignorant of Punjab's high rate of rural suicides for all these years.

Is it likely that around 60,000 people in Punjab may have committed suicide? Could the government be unaware of it? It is obvious that the government does not wish to acknowledge suicides.

WHY VILLAGERS COMMIT SUICIDE

An outline of the factors that drive Punjab villagers to take their own lives is given under 'Debt Trap' (refer to page 99).

While debt is the main, immediate provocation to suicide, other factors also contribute.

An important contributing cause is the break-up of the joint family system. Brothers marry and set-up their separate households. Land is divided among the brothers and with each succeeding generation, the holdings shrink. Often division of land creates bitterness between the brothers. Even if each brother gets exactly the same measure of land, land quality varies, so one brother may feel that the land parcelled out to him is inferior in fertility or in access to water. This is a very old story in Punjab. In fact, in the tale of *Heer–Ranjha*, the centuries-old folk romance, the hero, Ranjha, is the youngest of many brothers and his father's favourite son. When the father dies, the jealous elder brothers, apportion Ranjha his share of land, giving him land that is barren and uncultivable.

Another contributing cause is rising aspirations. Even in rural areas, people are exposed to television and see a glittering world of limitless consumerism. Because it is now easier to visit towns and cities, rural people also want to enjoy the affluence and high standard of living that prevails in the urban areas.

Lack of adequate institutional credit is another factor contributing to the rise in suicide rates. Farmers and farm labourers fall into the clutches of village moneylenders because they do not have access to formal banking institutions or credit cooperatives. If a farmer's assets are below a certain limit, a bank or credit cooperative will not extend a loan to him. The interest rate on an institutional agricultural loan is between 9 to 15 per cent and if he must turn to a moneylender, he will have to pay an interest ranging between 27 to 60 per cent. Moneylenders charge on the basis of every ₹100 advanced—for instance ₹2 per month on every ₹100 or ₹6 per month on every ₹100. Every 6 months, the balance is calculated. Sometimes, they advance ₹100 and ask to be repaid ₹1 every day.

Multiple Suicides

In some cases three or four members of the family commit suicide one after the other. Often it happens that the deceased would have managed to conceal his despair and the suicide comes as a surprise to the family and neighbours, but this is true only when the suicide is an isolated event. Once a suicide has taken place in a family, observers are 'tipped off' that the family is under heavy stress and they suspect that other members of the family may also be driven to the same extreme step. Between the first and subsequent suicide, there is generally enough time lapse for the government to provide relief. It is in the absence of this relief that despair sets in and leads to more suicides. Had the Government of Punjab provided relief to such families, as was done in the south Indian states, multiple suicides could have been prevented.

Tables 9.1, 9.2 and 9.3 list the cases of multiple suicides from three adjoining villages—Bakhora Kalan, Chottian and Balran, respectively.

Table 9.1: Village Bhakhora Kalan multiple suicides

Year	Date	Victim	Son of	Occupation	Age	Cause of death	Land holding	Debt		
1992	1 July	Gurmail Singh	Bant Singh	Agricultural labourer	19	Under train	Nil	25,000	PA	Two suicides: Baljinder and Gurmail (brothers)
1994	6 October	Baljinder Singh	Bant Singh	Agricultural labourer	18	Under train	Nil	40 000	PA	
1994	17 August	Lakhwinder Singh	Gurdip Singh	Agriculturist	30	Hanging	sold all	100,300	PA	Two Suicides: Kulwinder and Lakhwinder (brothers)
1995		Kulwinder Singh	Gurdip Singh	Agriculturist	26	Hanging	7 acre	500,000	PA	
1994	16 July	Naib Singh	Jhanda Singh	Agriculturist	25	Poison	sold all	50,000	PA	Two suicides: Brother Nek Singh committed suicide in 1986
1997	20 July	Rani Kaur	W/o Nirmal Singh	Agricultural labourer	18	Poison	Nil	53,000	PA	Two suicides: Rani Kaur and Ajaib (mother and son)
2000	24 September	Ajaib Singh	Nirmal Singh	Agricultural labourer	25	Poison	Nil	90,000	PA	
2000	15 December	Satpal	Dhag Singh	Agriculturist	25	Hanging	5 acre	200,000		Two suicides: Satpal and Mihan (brothers)
2005	12 October	Mihan/Mehr	Dhag Singh	Agriculturist	35	Hanging	1.5 acre	250,000	PA	
2005	20 May	Ramphal	Raghbir Singh		0					Two suicides: Brother's wife's suicide (Jagtar s/o Raghbir Singh)
2006	12 December	Malkit/Rani Kaur	W/o Jagtar Singh	Agriculturist	23	Hanging	2.5 acre	250,000	PA	

Source: Author.

Table 9.2: Village Balran multiple suicides

Year	Date	Victim	Son of	Occupation	Age	Cause of death	Land holding	Debt		
1989	17 February	Mishra Singh	Gajjan Singh	Agriculturist	38	Poison	1.5 acre	200,000	PA	Two suicides: Mishra and Harnek
1989	17 February	Harnek Singh	Gajjan Singh	Agriculturist	24	Poison	1 acre	132,000	PA	
1990	19 August	Sukhdev Singh	Bachan Singh	Agriculturist	32	Poison	2 acre	150,000	PA	Three suicides: Sukhdev, Sansi and Gama (brothers)
1992	13 October	Sansi Singh	Bachan Singh	Agriculturist	25	Hanging	2 acre	300,000	PA	
1992	1 January	Gama Singh	Bachan Singh	Agriculturist	40	Poison	4 acre	200,000	PA	
1988	7 June	Nikka Singh	Gurdev Singh	Agricultural labourer	25	Poison	Nil	80,000	PA	Four suicides: Darshan, Nikka and Nachhattar.
1993	14 April	Nachhattar Singh	Gurdev Singh	Agricultural labourer	40	Poison	Nil	80,000	PA	Krishan, s/o Nachhatar Singh
2001	1 November	Krishan Singh	Nachattar Singh	Agricultural labourer	19	Poison	Nil	25,000	PA	
2003	9 November	Darshan Singh	Gurdev Singh	Agriculturist	25	Poison	Nil	70,000	PA	

Year	Name	Father's Name	Occupation	Age	Method	Land	Amount		Remarks
1993	Aaloo Singh	Saon Singh	Agricultural labourer	30	Poison	Nil	50,000	PA	Three suicides: Aaloo, Kala and Geja (brothers) s/o Saon Singh
1996	Geja	Saon Singh	Agricultural labourer	22	Poison	Nil	60,000	PA	
2003	Kala Singh	Saon Singh	Agricultural labourer	26	Poison	Nil	150,000	PA	
1995	Dharam Das	Bhajan Das	Agricultural labourer	40	Hanging	Nil	70,000	PA	Two suicides: Rampal and Dharam Das (brothers)
2004	Rampal	Bhajan Dass	Agricultural labourer	30	Poison	Nil	40,000	PA	
1991	Nachhattar	Gamdoor	Agriculturist	19	Poison	1.5 acre	200,000	PA	Four suicides: Nachhattar, Darshan, Kaka and Jarnail (brothers)
1995	Darshan Singh	Gamdoor	Agriculturist	45	Poison	2 acre	150,000	PA	
1997	Kaka Singh	Gamdoor	Agricultural labourer	40	Poison	Nil	150,000	PA	
2004	Jarnail Singh	Gamdoor	Agriculturist	40	Poison	2 acre	200,000	PA	
1998	Sadhu/Kaku Singh	Bachna Singh	Agriculturist	51	Poison	2 acre	200,000	PA	Two suicides: Sadhu and Sibu (brothers)
1998	Sibu Singh	Bachna Singh	Agriculturist	35	Poison	4 acre	200,000	PA	
1998	Akki Kaur	Gurmail Saini	Agricultural labourer	18	Poison	Nil	200,000	PA	Two suicides: Avtar and Akki (brother and sister)
2005	Avtar Singh	Gurmail Saini	Agricultural labourer	25	Poison	Nil	300,000	PA	

Source: Author.

Table 9.3: Village Chottian multiple suicides

Year	Date	Victim	Son of	Occupation	Age	Cause of death	Land holding	Debt	Panchayat affadavit	Other suicides in the same family
1991	24 September	Pala Singh	Maghar Singh	Agriculturist	21	Poison	2 acre	150,000	PA	Two suicides: Gulzari and Pala (brothers)
2002	1 March	Gulzari Singh	Maghar Singh	Agriculturist	35	Poison	sold all	150,000	PA	
1993	9 August	Dulla Singh	Mohinder Singh	Agricultural labourer	30	Hanging	2 acre	76,000	PA	Two suicides: Dulla and Bhatti (brothers)
2000	1 January	Bhatti	Mohinder Singh	Agriculturist	35	Hanging	5 acre	500,000	PA	
1993	1 July	Natha Singh	Nand Singh	Agriculturist	45	Poison	5 acre	300,000	PA	Two suicides: Natha and Budh (brothers)
1993	15 May	Budh Singh	Nand Singh	Agriculturist	60	Under train	3 acre	250,000	PA	
1997	6 May	Bhola Singh	Jagga Singh	Agricultural labourer	35	Under train	Nil	100,000	PA	Two suicides: Balla and Bhola (brothers)
2007	1 February	Balla Singh	Jagga Singh	Agricultural labourer	24	Electrocution	Nil	70,000	PA	

Source: Author.

MISSING PERSONS

A new phenomenon, as yet not understood, is the disappearance of men from their villages. This is being observed, not only in Punjab villages in the Malwa area but in Haryana villages as well. The missing men are of all ages and some are reported to have disappeared long ago.

When a man goes missing there may be several plausible explanations:

1. He has fled from the demands of his creditors and the responsibilities that his family places upon him.
2. He has been murdered and those who have done him to death have concealed or destroyed the body.
3. He has met with an accidental death in a place far from his home. If he had carried no identification, his body might have been disposed of by the police as unclaimed.
4. He has committed suicide at a place far from his home. Either the body could not be recognised or he threw himself in a canal where the body would be washed away.

As per the law, when a person is missing for seven years, the heirs may submit a petition with proofs to the deputy commissioner of the district, asking that the person be declared dead. In any case, seven years is a long time for the family to wait for the division of the property and assets of the dead person.

In some respects, missing men are more of a problem to their families than men who have committed suicide. If the person still has some landholding or if the family home also stands in his name, it cannot be legally transferred to his heir. If a widow or child wants to apply to a loan against the agricultural land or house, no formal credit institution can grant such a loan to a person who is not the legal owner of the asset.

The wife, aged parents and children cannot apply for state pensions as it is uncertain whether the breadwinner is alive or dead.

A missing person is automatically the concern of the police. Police involvement means harassment at the very least and a high probability of extortion. Sometimes, the police knock on the door with the story—maybe concocted by themselves or supplied by some ill-wisher—that the missing man fled after committing a theft or some other crime. Or it may be alleged that the missing man has been murdered by a relative. No matter what the story is, police involvement is bad news.

On 4 January 2004, Prakash Singh, son of Sita Singh, a farmer of village Balran, took his wife to a nearby *Mandi* (Lehragaga) and purchased rations for the house for a month. He sent her home assuring that he would join her later. He never did. Days later his body was recovered by the police from a distant village pond. According to police, the body bore no marks of injury, suggesting the death to be a case of suicide. Other such cases are mentioned in Table 9.4.

Table 9.4:   **Missing persons in village Balran**

Name	Father's name	Age	Occupation	Missing since	Remarks
Chimna Singh	Bachna Chowkidar	40	Agricultural labourer	7 years	Missing
Gurdial Singh	Mehar Singh	70	Agricultural labourer	6 years	Missing
Gurmail Singh	Niranjan Singh		Agricultural labourer	8 years	Missing
Gurjant Singh	Mohinder Singh	22	Agricultural labourer	5 years	Missing
Meli singh	Jaila singh	35	Agricultural labourer	4 years	Missing
Makhan singh	Mishra singh	30	Agricultural labourer	10 years	Missing
Mewa Singh	Matu Ram	40	Agricultural labourer	5 years	Missing
Saon Singh	Kesar Singh	53	Agriculturist	2 years	Missing
Bhola Singh	Dharam Singh	20	Agriculturist	7 years	Missing
Parkash Singh	Sita Singh	30	Agriculturist	5 days	Drowned in a distant village pond. Body found with no injury marks.
Janta Singh	Mohinder Singh	19	Agriculturist	8 years	Missing
Kulwant Singh	Dhyan Singh	18	Agriculturist	10 years	Missing

Source: Author.

DROWNED PERSONS

For many years, MASR has repeatedly brought to the government's notice the problem of a large number of bodies seen floating down the canals. Punjab has almost 5,000 kilometres' of canal network. The phenomenon has also been highlighted in the press.

During the devastating floods of 1988, Bhakra Dam sluice gates were opened, ostensibly to save the dam, resulting in massive floods downstream. Punjab declared the recovery of 900 bodies from the floods. Pakistan reported sightings of 1,700 bodies. It takes 24 hours for a body to surface. Many must have gone unobserved this way and some may have floated down during the night. Punjab statistics stuck to the 900 figure. The Punjab government declared that crops and infrastructure worth ₹270 billion was destroyed. The central government granted some amount for rebuilding infrastructure but gave not a single rupee for crop loss.

The Punjab government is deliberately 'careless' when it comes to recording unnatural deaths. While many of those whose bodies were found floating down the canals during 1984–94 would have been the result of custodial killing of Sikhs suspected of being militants, those from 1994 onwards may represent a large percentage of farmer and farm labour suicide. Considering that there is no let up in the downward skid of rural economy, we may expect the number of agro-related suicide bodies to increase in the canals of Punjab.

The police take no cognisance of suicides and the victim's kins are reluctant to report the cases to the police. MASR believes that the police is actively discouraging people from reporting such cases. The police approach is clearly illustrated by the their instructions to canal *beldars*[4] not to remove a body from the canal but to push it downstream.[5]

The gruesome use to which the Sirhind Canal was put during the period of militancy was reported in 1992 (27 March) in *The Pioneer*. Correspondent Navin Grewal reported that the police were dumping bodies of suspected militants and their sympathisers killed in custody,

---

[4]A *beldar* is an employee of the State Irrigation Department. He is assigned duties for a certain stretch of the canal and he is responsible for reporting any defect in the canal or any illegal use (theft) of water.
[5]Balwant Garg, 'Can't Bury It, Push It Away', *The Times of India*, 9 July 2001.

in the canal. He visited the canal—covering the distance from the Kotla branch of the Sirhind Canal from the Moharana bridge on the Nabha-Malerkotla road to the Babanpur bridge on the Dhuri-Malerkotla road, a distance of about 8–10 kilometres—when repairs were in progress and release of water from the Bhakra Dam had been stopped, allowing bodies lying on the bottom to be seen. He reported that at least seven bodies were pulled out from the canal in this short stretch. To further authenticate the findings, the Bathinda branch of the same canal was checked out. The facts were similar. Both these canals pass through the Ropar, Ludhiana and Sangrur districts. Some 12 bodies of young Sikh boys were found, some with their hands and feet tied. All the bodies were those of Sikh youth.

He connected the discovery of so many bodies to the impending Punjab elections and said that many people had disappeared. He also mentioned that a police interrogation centre was located on the canal a little ways upstream. He credited MASR, Punjab Human Rights Organisation (PHRO) and the Peoples' Union of Civil Liberties (PUCL) for bringing the recovery of the bodies to the notice of Prime Minister P.V. Narasimha Rao in a letter.

Both these canals pass through the Ropar, Ludhiana and Sangrur districts, which were at that time the main militancy-affected districts; even more than Amritsar and Gurdaspur.

In his report, Grewal also observed the indifference of the police, accusing them of ignoring bodies even when brought to their attention. He said that they preferred to let a body float by, passing on the responsibility to the next police station. He learnt that many bodies thrown into various branches of the canal finally reach other states and referred to an admission by senior Punjab government officials that the Rajasthan government had complained to the Punjab government and expressed serious concern over the increase in the number of bodies flowing into that state.through the Rajasthan and the Abohar Canals.

The situation was unchanged 16 years later when an *Indian Express* correspondent visited the siphon net on the Bhakra Canal at Khanauri. His report brought out the neglect of the drowned dead.[6]

[6]Aman Sood, 'On Punjab–Haryana Border, Bodies Pile up along Canal', *Indian Express*, 15 September 2008.

He reported that bodies floating in the water at this place were a common sight (about 20 every month according to a local resident) and although the police station was nearby, the police scrupulously avoided taking note of the bodies so as to spare themselves time and trouble. He quoted the then Khanauri Station House Officer (SHO) Hardeep Singh, who at first claimed that the area was not in his jurisdiction and denied knowledge of the corpses. When the correspondent pointed out that the canal was a few hundred metres from his police station, he admitted to the long-drawn procedure and legal tangle after fishing out the bodies with no one to claim them. Were the police to retrieve a body, they would bear the responsibility for cremating the dead, lodging a First Information Report (FIR) and settling jurisdiction issues between Punjab and Haryana. A diver, Dharminder Singh, who had been bringing out bodies for the last 15 years, told him that there is no rule or law on cremating such bodies. The police is also reported to have issued verbal instructions to the State Irrigation Department canal *beldar*s and local police not to take bodies out of the canal but to push them downstream. This resulted in a large number of bodies floating into Rajasthan. The Rajasthan Chief Secretary repeatedly contacted the Government of Punjab to protest, as the bodies eventually reached Rajasthan.

Like Grewal, Sood also referred to repeated protests by the Rajasthan government to the Government of Punjab regarding the bodies.

In January 2011, the *Hindustan Times* (3 January, Chandigarh) correspondent visited Khanauri and reported that bodies were still seen floating down the canal. Seven bodies were spotted on the day of his visit. There is no improvement in the situation but by this time, it can be surmised that most of the bodies belonged to suicide victims with perhaps a few victims of murder.

ATTEMPTED SUICIDE CASES

Many persons attempt suicide unsuccessfully. The victim's family rushes them to private doctors. Legally, the doctors are required to report attempted suicides but doing so would involve the doctors and the families in police investigations. These cases go unreported.

Suicide survivors are worth interviewing as they can speak about the exact circumstances that drove them to make an attempt on their own life. These suicide survivors are also in need of counselling, assistance and rehabilitation. Their number is large enough to merit government concern. Some of them have become physically incapacitated and are unable to do any physical work.

PSYCHOLOGICAL TRAUMA

Cases of persons who have lost their mental balance under extreme economic stress are too numerous to count and these people too need care. In many cases, the victim develops severe mental symptoms, typically manifesting as complete withdrawal. When men—the family breadwinners—are afflicted in this manner, it means that the entire family is under extreme stress. For one thing, they have no source of income and for another it is very difficult to look after members of the family who have lost their mental balance. In terms of suffering for the family, this situation is as bad as or worse than losing a member to suicide. Such victims should be considered equally deserving of rehabilitation, including medical treatment and compensation.

Extreme psychological stress affects every human being. Moreover, women are more vulnerable to mental trauma as they remain largely house-bound and have few outlets or other diversions. They tend to continually brood over the family's difficulties.

WHY SUICIDES ARE MORE IN SOME YEARS, LESS IN OTHERS

In 2005 and 2006, the number of rural suicides decreased slightly on account of substantial escalation in land prices. By selling land, the farmers were able to repay all or most of their debts. But by 2007, suicides were again showing an upward trend. This trend can only accelerate as landholdings gradually shrink.

In 1971, land ceiling legislation imposed an upper limit of 17.5 standard acres on holdings of agriculture land. As sons followed fathers, the holdings were divided and re-divided among each succeeding generation. At present about 70 per cent of farms are below 5 acres.

For the past many years, farmers cultivating 10 acres or less have not been able to make a profit. It is just not possible, given the present price structure, to survive on less than 10 acres. Already 30 to 35 per cent of those who once cultivated their own land have sold out all that they owned and have become farm labourers.

Farmers are opting out of owning land in two stages. In the first stage, they cultivate their own land and add more land on contract basis for cultivation. In the second stage, they sell all their land and begin working for wages on the fields of others. This reduced status is highly humiliating to them; so he prefers not to work for the landowners in their own villages but instead goes to a distant village to work. Gradually they accept work in their home villages.

BREAKING THE SILENCE

Accurate data collection of suicides generally takes two to three visits to a village as family members are generally reluctant to divulge this information to an outsider. The neighbours in deference to the sensitivity of the affected families also try to conceal such suicides. The fear of police harassment and extortion is a strong consideration for withholding information. It is only lately that realisation has set in that there may be a possibility of receiving compensation for the next of kin from the government. This has nudged families somewhat to own up to suicides.

However, government servants' one-time visits to collect suicide data does not yield satisfactory results because:

1. The visits are under publicised.
2. The government's visiting teams are not punctual and the assembled villagers tend to disperse before the team arises.
3. The loss of work and wage discourage people from attending.

4. Fear of the police presence with the investigating teams' acts as
   a deterrent.

The methodology adopted by the late Aman Sidhu of verifying
suicide data through the Panchayats and getting the Gram Panchayats
to give affidavits confirming that the suicide was due to economic
hardship and debt, has been accepted by the National Commission of
Farmers as being the most viable and reliable method.

IRRIGATION AND SUICIDE

Along with the Green Revolution came the diversion of Punjab's river
waters to the non-riparian state of Rajasthan. In the Indus Water
Treaty, India had wrongly projected Rajasthan as a much larger
consumer of Punjab river water, although the Ganganagar area of
Bikaner state in Rajasthan was given only a small supply of Punjab's
river waters through the Gang Canal on payment of royalty. Punjab
retained the right to stop the supply if it so desired.

Punjab was assured that as Rajasthan was a desert area it would
not be able to utilise much of the water allocated to it. Therefore,
most of the water would revert back to Punjab.

If, at that time, Punjab had protested against this allocation to
Rajasthan from its three eastern rivers, then India's share of water
would have been substantially reduced; however, of this reduced
quantity apportioned to India, that water would have been made
available to Punjab alone.

The extent of the reduction of canal water supply to Punjab
farmers can be illustrated by my personal experience. Up to 1983,
canal water supply to my farm in Moonak subdivision was sufficient
to irrigate between 18 to 20 acres in a span of 24 hours. Around 1983,
the canal channels were paved with concrete to prevent seepage and
the diametre of outlets was reduced. This cut the flow of water so that
it sufficed for only 10 to 12 acres to be irrigated in the same time span.
By the end of the 1980s, water supply in the canals became irregular as
frequent cuts were imposed. We are now able to irrigate on an average
about 6 acres of land. The canal tail end farmers are the worst affected.

THREE DISTINCT ZONES OF MOONAK AND LEHRA SUBDIVISIONS

The area of Moonak subdivision, now divided into two subdivisions of Moonak and Lehra, comprises the following three distinct zones (Table 9.5):

1. Canal tail end villages
2. Villages along the Ghaggar river
3. Villages bordering Haryana

Ten contiguous villages have been taken from each category to study the level of suicides.

Because of frequent floods, the flood-prone villages' soil quality is good and requires less fertiliser. The quality of water is better than other zones and the cost of extraction of water is less as the water table is much higher. These villages register fewer suicides than the other two zones because the economic needs of these villages are tuned to a single crop and people find survival easier than in the other two zones.

A similar situation is observed with regard to those villages where refugee population from West Pakistan has displaced the Muslim population in 1947, who in turn has gone to West Pakistan. These refugee farmers from West Pakistan had to work harder to rehabilitate

Table 9.5: Lehra and Moonak zones

Lehra & Moonak zones	Population	Suicides	Percentage
Zone 1: Canal tail end villages: Chollian, Chural, Bakhora, Balran, Lehal Kalan, Lehal Khurd, Salemgarh, Hamirgarh, Dudian and Gobindpura Jawaharwala	34,634	367	1.06
Zone 2: Ghaggar river flood prone villages: Benarsi, Baopur, Kauagaon, Holipur, Theri, Chandu, Mardui, Banga, Bushera and Makror	23,598	142	0.60
Zone 3: Villages bordering Haryana (Ghaggar food unaffected): Khanaui Khurd, Chalha Gobindpura, Karoda, Gulhari, Thaska Bhoolan, Bahminiwala, Ramgarh Gujran, Mariana and Khokhar	20,614	238	1.15

Source: Author.
Note: Ghaggar villages are few, whereas the largest number comprise the canal tail end villages.

themselves. This habit continues and they find themselves in a better position to face agro-desperation.

After the 1988 floods in Punjab, which devastated the Ghaggar villages, along with the Advisor to the Punjab Government Mr Julio Ribeiro, I visited the Ghaggar-flooded villages. Through Advisor Ribeiro, we approached the Punjab Governor for two relief measures. In principle, he accepted both these demands and assured us of his support in getting them through and advised us to make a formal representation of these two demands to the government which we did.

WHAT RESIDENTS OF THE FLOOD-PRONE VILLAGES WANT

The resident of the flood-prone villages want the following:

1. Build *bandh*s (levee) on both sides of the river Ghaggar from Khanauri Kalan to Jakhal.
2. Compensation for damage to crop by floods. This has been denied on the ground that the villages are historically flood-prone.

With regard to the first demand, the Secretary of Irrigation and Power, Government of Punjab, wrote to us informing us that the demand has been accepted and the project was being forwarded to the central government for approval.[7]

Swarn Singh Boparai, KC, IAS, Secretary, Department of Irrigation & Power,
Government of Punjab
D.O. No. 1/29/88-IPW(3) 17622 40184, 1/8/90.

Chandigarh, the 30th July, 90.

My dear Sardar Sahib,

Please refer to your note presented to me regarding construction of Bandh along river Ghaggar. I have got the matter looked into

---

[7]This letter is reproduced to show the level of dependence of the state government on the central government. It may be noted that ₹60 lakh is barely enough to build a medium-sized house in these times.

by the Chief Engineer/Drainage, Irrigation Works, Punjab. An agenda note of the scheme for the construction of bundh on both sides along river Ghaggar from Khanauri Acquaduct under B.M.L. to Haryana border near Jakhal in Punjab area amounting to ₹1529 lacs has been formulated by the Drainage Administration. The scheme is to be recommended by the State Technical Advisory Committee and then approved by the State Flood Control Board. The scheme, thereafter, will be sent to the Planning Commission through the Central Water Commission after getting clearance from the Ghaggar Technical Committee. **Since the scheme has inter-state aspects and costs more than ₹60 lacs besides its clearance from the Central Government agencies, as such it is likely to take some time. However all out efforts will be made to get the scheme implemented expeditiously.**

With regards,

Yours sincerely
Sd/-
Swarn Singh Boparai

The second demand was not accepted and we were compelled to take the matter to the Supreme Court. Due to prolonged litigation, we later made a direct appeal to the Chief Minister of Punjab, S. Harcharan Singh Brar, explaining to him why compensation should be provided to the flood-prone villages of river Ghaggar (the letter is reproduced below), which resulted in frequent damage to their summer crop brought about by disturbing the free flow of the river through numerous misguided development schemes of the government.[8]

16 June 1996
Hon'ble S. Harcharan Singh Brar, Chief Minister, Punjab.
Chandigarh
Sub: Ghaggar Bund

---

[8]S. Harcharan Singh Brar, being a farmer himself, appreciated the problem and not only extended flood relief to these villages but enlarged compensation for the flooded area from 5 acres to 10 acres. It is this factor that has been largely responsible for reduction in the level of suicides in the Ghaggar-belt zone.

Dear Mr Brar

In September 1988 when the Ghaggar flooded vast areas of southern Punjab and Haryana, Mr. J.F. Ribeiro, the then adviser to the Punjab Government visited the flood ravaged area of Sub-Division Munak, district Sangrur and while addressing the villagers at Munak, and later at village Bushera, announced compensation at the rate of 700/- per acre upto 5 acres. Though Government later went back on this assurance, he did help to set rolling our demand for Ghaggar river bund.

A demand was made by the undersigned on behalf of the flood affected villages for the construction of two sided Ghaggar bund covering the river bed from Khanauri Mandi to Jakhal Mandi. The financial implication of this demand was briefly discussed and Mr Ribeiro advised me to send the proposal personally to him for consideration by the Government.

The proposal was later sent to Mr. Ribeiro with a copy to Secretary Irrigation. The Secretary Irrigation confirmed that the proposal was examined by the Chief Engineer Drainage, Irrigation Works and a general note on this scheme, for the construction of bund on both sides on river Ghaggar from Khanauri to Jakhal near Haryana boarder amounting to ₹1529 Lac, was formulated by the Drainage Committee and then approved by the state Flood Control Board. The scheme, thereafter, was to be sent to the Planning Commission through the central water commission after getting clearance from the Ghaggar Technical Committee. However, the Secretary Irrigation assured that 'all out efforts would be made to get the scheme through expeditiously'.

After the 1993 floods Deputy Commissioner Sangrur, formally announced that he had received information from the Government that the two-sided Ghaggar bund had been sanctioned at a cost of 35 Crores and the construction work would start shortly. This information was also carried in some newspapers. Later we were informed that as the Government did not have sufficient funds, the project had been delayed. We are still awaiting Government's allocation of funds for this vital project.

In the past, river Ghaggar flooded the embankments in a cycle of 9 to 11 years, After construction of the Bhakra canal the floods became more frequent and after construction of the

S.Y.L. canal, floods have become an annual feature, sometimes flooding 28–35 villages 2 to 3 times in a single year.

Formerly the flood water used to drain out within a week, now it takes 3 to 6 weeks for the water to drain out. This is due to sudden increased flow of water coupled with insufficient siphon capacity at Chandpura aquaduct in Haryana, where the Bhakra canal crosses the Ghaggar River. In addition a number of kacha illegal bunds have been built down stream between Chandpura and Jakhal which obstruct the natural flow of water.

Haryana has also constructed, without prior approval of Punjab, a lake at Mustafabad near Kurukshetra. This releases surplus water during the flood season which has an adverse effect down stream in Munak Sub-Division.

A survey was conducted by the State Government in 1925 to determine the flood area that was affected by Ghaggar floods in Sub-Division Munak. The affected area was very small. From this it is obvious that magnitude of flooded area only increased after construction of these two canals i.e. Bhakra and S.Y.L. This has converted the flood prone land of Sub-Division Munak into a single crop zone thereby depriving the farmers the advantage of multiple cropping, This is, therefore clearly a man made catastrophe.

Whenever dams are built like the Bhakra, the Pong, the Salal, the Narmada, or changes made affecting the free flow of rivers the affected people are suitably compensated by providing alternate land sites plus help for incidental expenses. The farmers of these 35 now flood affected villages should be given half the land cost which at the prevailing rate is 2.5 to 3.00 lakh per acre, as they now get only a single crop from this land.

As a farmer yourself you are no doubt aware that a loss of a single crop financially cripples the farmer for 2 to 3 years. In the case of flood affected Ghaggar villagers of Sub-division Munak, their crops are ravaged by the floods every year. A point has been reached where distress sale of land is taking place and the farmers are so much burdened with heavy debts that cases of suicide have taken place.

I strongly urge you to come to the rescue of Ghaggar affected farmers of Sub-Division Munak and order the following measures to be taken:

1.   Immediate release of ₹35 crore for construction of Ghaggar bund. This would save crop area of 50,000 acres per year. The annual crop value of this works out to 40 crores.
2.   Give immediate Compensation to the flood affected farmers, at the rate of half the value of land as due to changes brought about by the construction of Bhakhra and S.Y.L. canals. They now get only a single crop from their fields.
3.   Electricity bills should be exempted for the period covering the damaged crop.
4.   Loans for agriculture inputs for the affected crop should be waived.
5.   Full compensation should be paid for crop loss, loss of houses, tube wells, Agriculture machinery and livestock.

With regards and Fateh,

Yours Sincerely

Sd/- Inderjit Singh Jaijee

CONSEQUENCES OF PAUPERISATION

1.   Abandonment of dependents such as aged parents and children; particularly girls are denied education as families cannot afford it.
2.   Joint families are growing increasingly rare in rural Punjab. When sons marry they ask for their share of the ancestral property. The family breaks up and the brothers are invariably unhappy with their share of the land. Even if the quantity of land received by each brother is exactly the same, one will feel that the quality of his particular parcel of land is less fertile than that of the other brothers. Poverty compels brothers to split up the parents with one looking after the father and another looking after the mother. Sometimes the bitterness between the brothers results in the parents being forbidden to meet each others' children and grandchildren.

3. Fragmentation plays a large role in pauperisation. As per Land Ceiling Act (1971), no one can own in excess of 18 standard acres. Punjab has had about 40 years of land ceiling. Suppose a family with only two sons, owned the maximum permissible 18 acres in 1971. By 1991, the sons are grown and married and demand their share, namely 9 acres each (the share would be 6 acres among the father and the sons and after death of the father, it would be 9 acres to each brother). By 2011, the sons own children are grown—let us say each has only two sons—so the holding of 9 acres is split and these young men receive 3 acres each. Even if the family has not gone into debt since 1971, the holding of the third generation must inevitably decline to 3 acres. This would be the case for the optimum permissible landholders, this category would not be more than 10–15 per cent in 1971. We can safely assume over 90 per cent of the farmers today are in marginal holding level of up to 2.5 acres. The land may stand in the name of father but de facto it is divided, government data not withstanding. The PAU economists have analysed farm profitability and concluded that holdings below 10 acres are unviable.

4. Consequences for women and girls:

    (a) Mothers occasionally desert the family, either through remarriage or return to natal family. Mothers deserting their children is a never-before-seen phenomenon in Punjab but it has an extenuating angle. As one mother told me, she left her children, knowing that the paternal grandparents would take them in and then the children would get more food. 'There was not enough food to go around so I decided to walk out so that the children could get a larger share', she said. She married her husband's brother, now a separate family. When food is shared, the mother invariably reduces her own portion so as to give more to the children.

    (b) Impoverishment means greater motivation for families to abort female foetuses, since families look toward the future and fear that they will be unable to raise, educate and marry a daughter. There has been an increase in the number of female infanticide cases. This is confirmed by latest census reports that show Punjab's sex ratio is at its lowest ebb.

(c)   In the villages of Punjab, when a woman is widowed she suffers not only emotionally but her social and economic status plunges. In most cases, poverty and debt were what compelled her husband to take his own life. When he is gone, the family's economic status can only worsen. Ideally the woman would remain with her husband's family who would support and protect her and her children. This ideal is rarely realised now. She might like to return to her natal family but if they had been able to help they would have come to their son-in-law's rescue long before. They may accept the daughter grudgingly and the arrangement rarely lasts for very long. In nearly all cases taken down in detail, the widows were on their own and forced to seek livelihood for themselves and their children. Some began doing domestic chores for better-off families in the village. For women who were once wives of farmers, this is humiliating; on the other hand it was less physically exhausting than other available employments such as field labour or road work. In any case, they simply did not have options and were forced to seek any type of occupation. Exploitation of these widows is common. Some are deceived by men who promise marriage. Some accept proposals of marriage from men whom they may never have considered suitable had their circumstances been better.

(d)   The purchase of brides from remote areas is another consequence. The better-off families from neighbouring villages refuse to give their daughters to men from economically distressed families or villages. These rejected men then buy women from other states. One of the reasons why families try to conceal suicide deaths is that they fear that no one will want to give daughters as brides to their sons or accept their daughters as brides. (In more prosperous times, families from Haryana villages preferred giving their daughters to Jats on the

Punjab side. The Punjabi farmers were better off than their Haryana counterparts. This marriage preference was also on account of work culture: as per social norms, women in Haryana carried out all the agricultural work except ploughing. In Punjab, women did not work in the fields.)

5. There has been a change in the cultural character of the village. Traditional farming castes are unable to continue farming. Because of the absence of institutional credit, farmers typically rely on *arthiya*s and commission agents to loan them money. They pledge their land as surety. When they cannot repay—as frequently happens—their land passes to creditors. These creditors belong to non-farming castes. Economic status is gradually replacing land and caste status as bases of social differentiation in the village. Impoverished Jats have no choice but to become labourers, bringing them to shared interest with the traditional labouring communities.

6. Monetisation/commodification is obliterating traditional ethics and practices. This has a bright side and a dark side. On the bright side, monetisation is weakening the caste system. On the dark side, bonds within the village community that once ensured that the relatively affluent had an obligation to ensure the survival of the less affluent have been largely forgotten. Formerly the agriculturalists (Jats) were the dominant group in the village and all other castes, including *bania*s, artisans and service castes, catered to them. The position of the rural Jats has weakened greatly and the *arthiya*, commission agent and the village trader have gone up in the village hierarchical order.

7. There is now an increasing friction between moneylenders and *arthiya*s, and farmers and farm labourers. Moneylenders resort to illegal collection means, especially in cases of dubious debts. Standing instructions forbid the police to intervene in matters such as debt and eviction, but they are known to ignore this directive and lend muscle-power to the moneylenders. In response, farmers are beginning to join hands to resist. Resistance too sometimes takes a criminal turn: thefts, dacoities and even murders are being reported.

A climate of lawlessness is developing and if not arrested in time, has the potential to erupt in social upheaval on a mass scale.

Small and marginal farmers have sunk far below the poverty line and many have lost what little land they had and have become landless. Caste-based assistance does not help them. All poor people, regardless of their caste, deserve state assistance.

Like people in cities, villagers are also exposed to modern media such as films and television. What they see makes them aspire to more affluent lifestyles. When they realise that their chances of actually ever enjoying such a life are very slim, it often demoralises them. On the other hand, their own traditional culture has waned under the impact of a consumerist society and pauperisation. One once heard happy songs in rural Punjab, now the only songs left are dirges.

Demoralisation has a stronger impact on those who were once prosperous but have been reduced to penury by debt. Punjab has been the granary of India and an example of rural prosperity. To be reduced to debt and stringency is hard indeed. Disintegration of social support institutions such as family, kinship, Panchayats, etc., have left the poor people in a state of all-round despair and helplessness. There is no one to share their misery with and they become loners in their last days.

Division and fragmentation of farmers' movements has led to a situation in which instead of finding a solution in a collective way, individuals are left to fend for themselves.

SOCIO-ECONOMIC CHARACTERISTICS OF SUICIDE VICTIMS

More than 70 per cent of the suicide victims belonged to small and marginal farmers and landless labourers. Studies clearly established that the percentage share of suicides goes down with increase in the size of landholdings.

Suicide in rural Punjab is a phenomenon of youth. Alarming as it is, it is indicative of there being something drastically wrong with the social situation in which the youth in rural Punjab is placed.

In our data, out of the farmer victims, 20 per cent were engaged in cultivation of wheat and cotton and 65 per cent were engaged in cultivation of wheat and paddy. Among small/marginal/landless victims, almost all were exclusively dependent on agriculture for a living. None was found to have a family member earning income as a government employee. Dairy, in some cases, was the only source of additional income.

Suicides in rural Punjab are predominately a male phenomenon —82 per cent of the victims are male.

The study by Institute of Development and Planning, Amritsar, Association for Democratic Rights (AFDR) and others have found that 83.5 per cent were Jat Sikhs, 8 per cent were Scheduled Castes, another 5 per cent were Brahmins and one per cent belonged to other communities. Thus, Jat Sikhs plus Scheduled Castes accounted for more than 91 per cent of victims. This was also conveyed to the central government by Punjab's Chief Minister Amarinder Singh vide Punjab Status Report 2004 on 8 October 2004.

The IDC studies have found an inverse relationship of suicide rate with the educational attainment—30 per cent were found to be illiterate, another 26 per cent were primary pass and another 21 per cent were found to have education up to middle level. In quite a few cases it was observed that the victim first took a swig of hard liquor before swallowing the pesticides. This was perhaps taken to soften the pain of death and make the bitter taste of poison palatable. It may be recorded that this practice was also followed by the Romans before crucifixion.

The IDC study found that 68 per cent of the victims were inclined to drink heavily with the majority of these being chronic alcoholics; 26 per cent of the victims were alleged to be addicted to narcotics. On the other hand, AFDR found that only 10 per cent of the victims were addicted to drugs or alcohol. A reason for this divergence could be that the IDC study was based on victims as per police record, whereas ADFR went beyond the police record. (Excise is almost the sole revenue earner of the state. The Punjab government has set up liquor vends on almost all street corners of towns and in all villages.)

SOCIAL PROFILE ANALYSIS OF RURAL SUICIDES AS REFLECTED THROUGH MASR DATA

The social profile of suicide victims as revealed through MASR figures highlight the following distinct features:

1. The youths are predominant among the farmers committing suicides.
2. Rural suicides are exclusively committed by families dependent on agriculture.
3. Marginal and small farmers are major victims of agrarian crises, indebtedness, and consequently they commit suicide.
4. The process of pauperisation is an actively ongoing process in rural Punjab which is creating a different and distressed survival and economic situation for the marginal and small farmer.
5. The incidence of suicides is also phenomenal among agricultural labourers, which is one of the unique features of agrarian suicides in Punjab.
6. Globalisation, Privatisation, Liberalisation (GPL) and the agrarian crisis have penetrated into the grassroots.

The MASR data of farmer's suicides cover the period of 1988 to 2008. In order to understand the incidence of suicide, it has been classified into four periods, namely, 1988–92, 1998–2002, 1993–97 and 2003–08. It will be seen that the total farmers' suicides committed during the entire period of 1988 to 2008 is 1,774 of which 9.15 per cent has taken place during 1988–92, 39.48 per cent during 1998–2002, 21.30 per cent during 1993–97 and 30.07 per cent from 2003 to 2008. The highest incidence of suicides in terms of percentage has been committed during 1998–2002, followed by 2003–08; the incidence of suicide is only 21.30 per cent during 1993–97. The highest incidence is thus during 1998–2002; high incidence is during 2003–08; moderate incidence during 1993–97 and low incidence during 1998–92 (see Table 9.6 and Figure 9.1).

The highest incidence was in 2000, 2001, 2002, 2003 and 2004. These years coincide with the period of adverse impact of GLP on

Table 9.6:   Year-wise suicides: Moonak subdivision(1988–2008)

Year	Total	Percentage
1988	31	1.74746336
1989	13	0.732807215
1990	45	2.536640361
1991	21	1.183765502
1992	32	1.803833145
1993	36	2.029312289
1994	56	3.156708005
1995	92	5.186020293
1996	72	4.058624577
1997	68	3.833145434
1998	140	7.891770011
1999	72	4.058624577
2000	168	9.470124014
2001	149	8.399098083
2002	151	8.511837655
2003	131	7.384441939
2004	145	8.17361894
2005	98	5.524239008
2006	73	4.114994363
2007	92	5.186020293
2008	89	5.016910936
Total	1,774	100

Source: Author.
Notes: 1. Suicides have been continuous every year.
2. The suicide rate is higher from 1995 onwards.
3. It has come down slightly during 2006–07.
4. Year 1997 is a critical cut-off as the suicide rate consistently picks up after that.
5. There has been no intervention by the state or central government, like giving relief packages to farmers.

Punjab farmers. The years 1995 and 1998 also saw spurts in suicides. Two factors impact the farmers' well-being: one is the government-declared Minimum Support Price (MSP) and the other is the weather as it impacts the harvest.

Figure 9.1:   Year-wise suicides: Moonak subdivision (1988–2008)

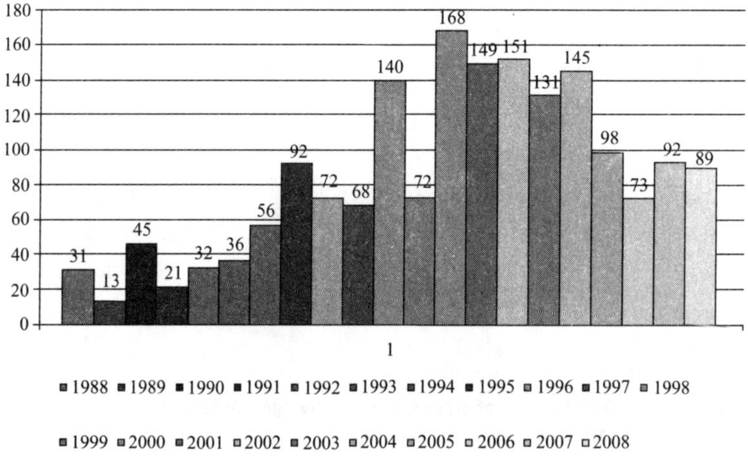

■ 1988 ■ 1989 ■ 1990 ■ 1991 ■ 1992 ■ 1993 ■ 1994 ■ 1995 ◫ 1996 ■ 1997 ◫ 1998

■ 1999 ◫ 2000 ■ 2001 ◻ 2002 ■ 2003 ◫ 2004 ◫ 2005 ◻ 2006 ◻ 2007 ◻ 2008

Source: Based on Table 9.6.

Age Group

In rural Punjab young people are more likely to take their own lives than older people. This is alarming and indicative of something drastically wrong with the social situation in which Punjab's rural youth is placed. The data presented in Table 9.8 is self explanatory as 55.62 per cent of the total suicides have been committed by persons aged between 15 and 30 years and another 33.57 per cent were committed by persons aged between 31 and 45 years. Only 10.37 per cent were committed by persons aged 46 years or older. This also shows the sensitivity of younger people to the issues of dignity as well as the drastic impact of indebtedness and agrarian crises on them.

The following points can be observed:

1. Incidence of suicide is higher among the younger-age group followed by middle-age group and lower among the higher-age group.
2. Younger-age group, both men and women, is most sensitive to economic distress.

Table 9.7: **Distribution of suicide victims by age: Moonak subdivision (1988–2008)**

Age (in years)	Suicides	Percentage
0–20	198	11.16121759
21–40	1,182	66.62908681
41–60	300	16.91093574
61–70	13	0.732807215
Unknown	81	4.565952649
Total	1,774	100

Source: Author.

Figure 9.2: **Distribution of suicide victims by age: Moonak subdivision (1988–2008)**

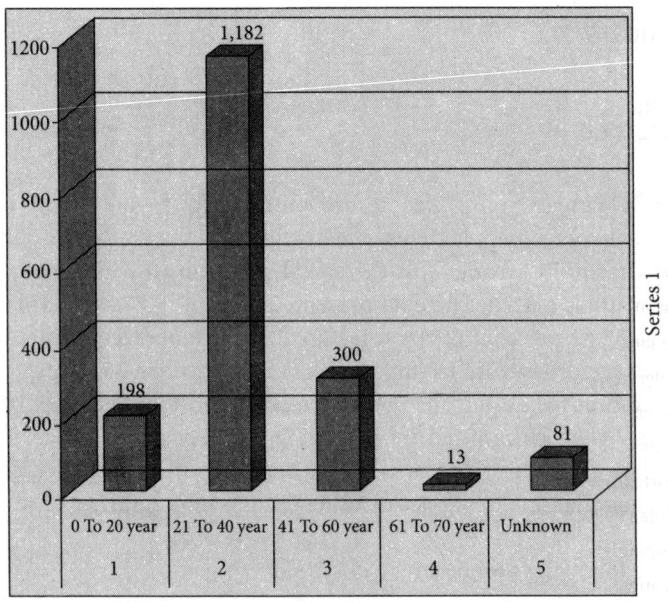

Source: Based on Table 9.7.

3. The causes of suicide have primarily been economic distress, indebtedness and humiliation, caused to the affected families by commission agents and institutional agencies.

4. The economic penury faced by the surviving nearest kin, particularly the widow and dependent children, is a matter of serious concern.

Means Adopted in Committing Suicide

Consuming pesticides has been the most frequent means adopted for committing suicide: 76 per cent of the total cases (see Table 9.8, Figure 9.3). This is followed by hanging. This is a specific symptom of the post-Green Revolution period. The other methods are the very crude forms of committing suicide, like throwing oneself under a running train, self-immolation and drowning. These reflect the desperation and distress of the victim.

Occupation

Of the total suicides, 60.97 per cent were committed by agriculturists, 38.33 per cent by agricultural labourers and 0.70 per cent by

Table 9.8: Distribution of suicides by means: Moonak subdivision (1988–2008)

Modes of suicide	No. of suicides	Percentage
Poison	1,276	71.9
Hanging	170	9.58
Under trains	93	5.24
Drowning	75	4.22
Burns	51	2.87
Heart attack	1	0.05
Sudden depression	2	0.11
Jumping	3	0.17
Attempted suicides	7	0.39
Electrocuting	19	1.07
Under trucks	1	0.05
Alcohol	5	0.28
Not known	71	4.0
Total	1,774	100

Source: Author.

Figure 9.3: Distribution of suicide victims by means: Moonak subdivision (1988–2008)

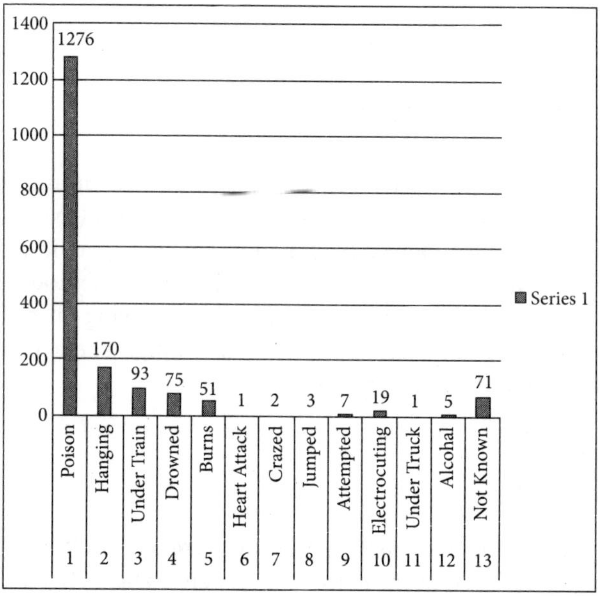

Source: Based on Table 9.8.

others (see Table 9.9, Figure 9.4). An overwhelming percentage of them were dependent on agriculture for their livelihood. Another important feature is that a substantial percentage of suicide cases are committed by agricultural labour, indicating concomitant effect on the class dependent on farmers for their survival. This is also one of the special features of suicides in Punjab as compared to other states.

Among the farmers committing suicide, 50.35 per cent are marginal farmers, 31.43 per cent are small farmers and 10.48 per cent are middle-class farmers. The predominance of marginal and small farmers in particular, reveals the impact of acute agrarian crisis on marginal and small farmers for whom agriculture is no more a viable proposition. Another grievous situation is that some of the farmers who have committed suicides have also lost their land. Thus, it is amply evident that survival and a distressed situation compelled the farmers to commit suicide. Most of the marginal and small farmers are selling

Table 9.9:   Distribution of suicide victims by occupation: Moonak subdivision (1988–2008)

Occupation	Suicides	Percentage
Agricultural labourer	684	38.55693
Agriculturist	1,078	60.76663
Service	5	0.281849
Shopkeeper	2	0.11274
Peon	1	0.05637
Cycle Repairer	1	0.05637
Trade	1	0.05637
Carpenter	1	0.05637
Tailor	1	0.05637
Total	1,774	100

Source: Author.

Notes: (i) Agriculturist and agricultural labourers are the two major groups committing suicide.

(ii) Punjab has the highest incidence of agricultural labourers committing suicides.

(iii) Agricultural labourers squarely depend on the informal institutions for debts, and so their cases for relief packages from the government need to be addressed on priority basis.

Figure 9.4:   Distribution of suicide victims by occupation: Moonak subdivision (1988–2008)

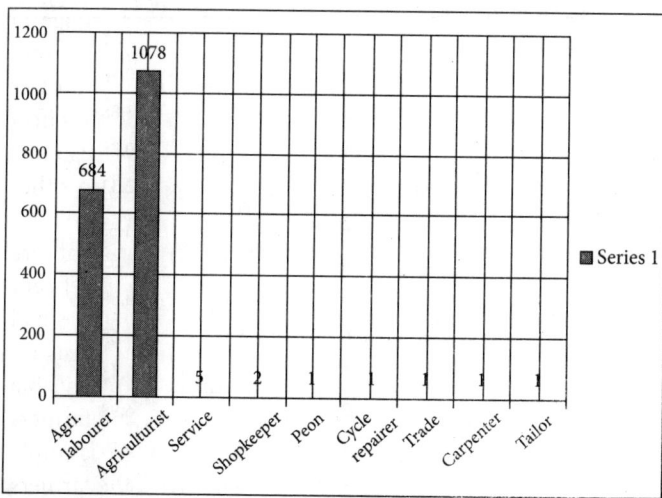

Source: Based on Table 9.9.

their land to clear the debts, indicating the accelerating process of marginalisation.

Women's Suicides

Although suicides in Punjab is predominantly a male phenomenon (see Tables 9.10 and 9.11 and Figures 9.5 and 9.6), women constitute 8.20 per cent of the total suicide cases.

Table 9.10: Distribution of suicide victims by gender: Moonak subdivision (1988–2008)

Male/Female	Suicides	Percentage
Male	1,599	90.13528749
Female	175	9.864712514
Total	1,774	100

Source: Author.

Table 9.11: Year-wise distribution of women suicide victims: Moonak subdivision (1988–2007)

Year	Females	Suicide incidence (in percentage)
1988	2	1.55
1989	1	0.78
1990	5	3.88
1991	1	0.78
1992	3	2.33
1993	2	1.55
1994	6	4.65
1995	5	3.88
1996	6	4.65
1997	7	5.43
1998	9	6.98
1999	4	3.10
2000	9	6.98

(Table 9.11 Continued)

(Table 9.11 Continued)

Year	Females	Suicide incidence (in percentage)
2001	11	8.53
2002	10	7.75
2003	8	6.20
2004	12	9.30
2005	9	6.98
2006	8	6.20
2007	11	8.53
Total	129	100

Source: Author.
Notes: 1. Women's suicides are related to their age and marital status.
2. Most of them belong to the younger age group, followed by the middle age group and lowest among the older age group, indicating highly sensitivity of youth.
3. Most of the women are married though the unmarried also constitute a minuscule part.
4. The reasons for committing suicides are distressed economic situation, emotive land issue and humiliation perpetrated on the family by the lending agencies.

Figure 9.5:   Distribution of suicide victims by gender: Moonak subdivision (1988–2008)

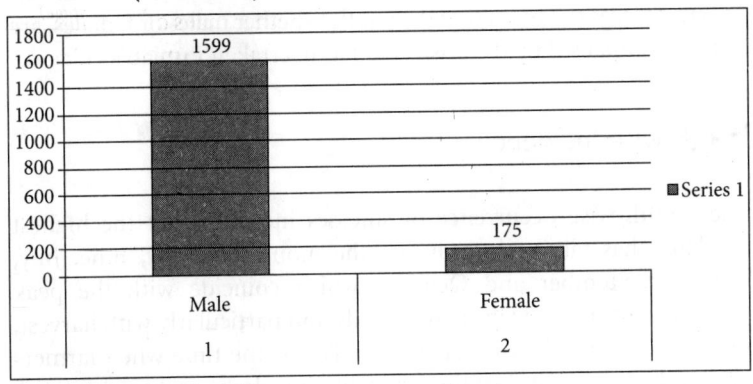

Source: Based on Table 9.10.

According to a study conducted by Professor Gopal Iyer of Panjab University, more women commit suicide in Punjab than in any other state. There is a positive correspondence between the periods of high

Figure 9.6:   Year-wise distribution of women suicide victims: Moonak subdivision (1988–2007)

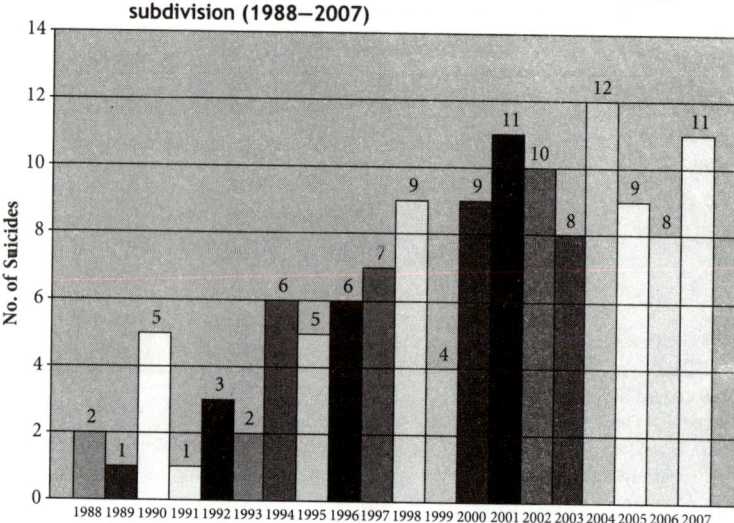

Source: Based on Table 9.11.

incidence of suicide committed by males and females. The women who committed suicides generally were young and although most of them were married, a few unmarried girls also took there own lives. This reinforces the point that the youth, whether males or females, are seriously impacted by the crisis that has overtaken Punjab's villages.

Month-wise Incidence

The month-wise occurrence of suicides indicates that the highest incidence has taken place during the months of May, June, July, August, September and October, which coincide with the peak agricultural season of wheat and paddy and particularly with harvest, sowing and transplanting operation. This is the time when farmers need cash not only to clear the past debts, but also to make payment to labourers to prepare for the next agricultural season. It also coincides with the time of loan recovery by institutional and informal lending agencies (see Table 9.12 and Figure 9.7).

Table 9.12:   Distribution of suicide victims by month of death: Moonak subdivision (1988—2008)

Sr.No.	Year	Suicides	Percentage
1	January	106	5.97519729
2	February	107	6.03156708
3	March	104	5.86245772
4	April	107	6.03156708
5 ·	May	178	10.0338219
6	June	172	9.69560316
7	July	191	10.7666291
8	August	190	10.7102593
9	September	154	8.68094701
10	October	137	7.72266065
11	November	108	6.08793687
12	December	84	4.73506201
13	Not Known	136	7.66629087
	Total	1,774	

Source: Author.

Figure 9.7:   Distribution of suicide victims by month of death: Moonak subdivision (1988—2008)

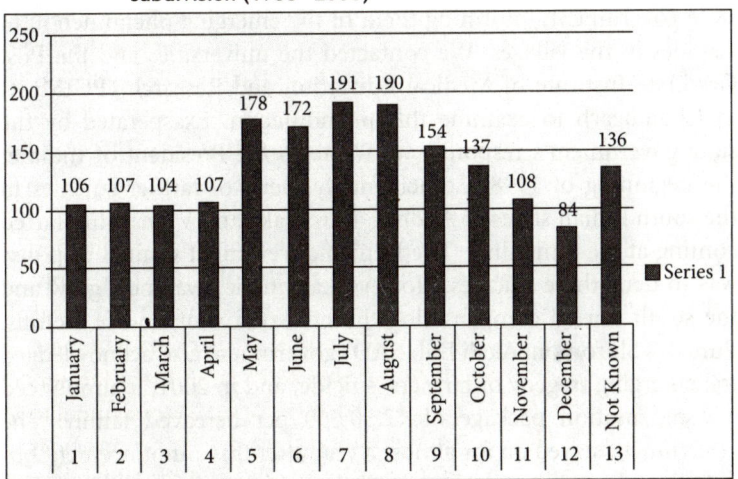

Source: Based on Table 9.12.

10

Rehabilitation and Compensation

RELIEF TO FARMERS AND FARM-LABOURERS

In 1990, MASR sent its first written report on rural suicides to the Health Secretary, Government of Punjab, apprising him of nine suicides in a single village of district Sangrur within a course of 2.5 years. This was followed up by a number of reports sent to the Deputy Commissioner of Sangrur and other senior officials of the state government, informing them of the emerging phenomenon of suicides in the villages. We contacted the universities and the Post Graduate Institute of Medical Education and Research (PGIMER) in Chandigarh to examine this phenomenon. Exasperated by the state government's response, we wrote to the President of India at the beginning of 1998. Coincidentally, news of farmer suicides in the south Indian states of Andhra, Karnataka and Vidharbha started coming at the same time. The Punjab government's initial response was to deny these suicides. However, as public awareness grew and the south started compensating the next of kin of suicide victims, Punjab's Shiromoni Akali Dal (SAD) government also acknowledged the emerging tragedy of farmers' suicide, and in 2001, it announced a compensation package of ₹2,50,000 per bereaved family. The government stayed on for almost a year after this announcement but failed to take any steps to honour its commitment.

The 2002 state election returned the Congress to power in Punjab and Captain Amarinder Singh became the Chief Minister of the state.

He had been sympathetic to the farmers' cause but when he failed to deliver, MASR filed a writ in the High Court seeking implementation of the Punjab government's unfulfilled assurance given to the next of kin of suicide victims. The case was vigorously fought by High Court advocate R.S. Bains.

While disposing of MASR's cases (Civil Writ Petition No. 1211 of 2005 and Civil Writ Petition No. 17844 of 2004),[1] the Hon'ble Punjab and Haryana High Court passed the following order:

> In the written statement filed on behalf of the Government of Punjab, it is stated that the scheme devised by the previous government for extending monetary assistance to the families of farmers who have committed suicide due to indebtedness has been found unworkable for various reasons, spelt out in the affidavit. One of the reasons mentioned there is that such a scheme could, in fact, encourage the committal of suicide in some cases. It is however stated that on 05-05-2005, the State Government has set up the Punjab Farmers' Commission to find out ways and means to rejuvenate Punjab's agrarian economy. The said Commission shall also look into the issue of suicides by the farmers in the state, its causes, and will suggest remedial measures including rehabilitation and compensation to the affected person/families.
>
> We feel that in view of the setting up of the Commission, no further orders are called for in this petition. Nevertheless, we are confident that the Commission shall go into the root cause of suicides by the farmers and submit its report expeditiously so that the follow up action, including payment of compensation to the affected families, if so recommended is taken promptly, otherwise the whole purpose of the constitution of the Commission shall get frustrated.

PUNJAB FARMERS COMMISSION

The Punjab Farmers Commission was accordingly set up by the state government and it recommended a number of measures to

---

[1][MASR through its convener, S. Inderjit Singh Jaijee], petitioners vs Shri G.S. Cheema and other respondents: application under Section 151 of CPC praying for dispensing with the filing of certified copies of annexures P-1 and P-2 and detailed affidavit of the petition.

ameliorate the desperate situation for Punjab's farming community, including rehabilitation and payment of relief to families of suicide victims. But relief to the next of kin of the suicides victims was not provided and nor were adequate measures taken to improve the position of the farm sector. MASR again approached the High Court in July 2006 in a Contempt Petition under Section 12 of the Contempt of Courts Act, 1971, praying for initiating proceedings against the respondents for wilful disobedience by the respondents of the High Court's orders.

The petition was heard on 9 December 2006. The Punjab government drew the ire of the court for avoiding a categorical stance on implementing recommendations of the Punjab Farmers Commission, including directing commission agents to pay farmers by cheque. (As of now and despite the state government's assurance to the court that payment by cheque would be enforced, this practice is still not followed. Payment by cheque would mean that commission agents would have to pay farmers the price of their produce first and afterwards recover any dues, rather than deducting dues first and then paying farmers the balance amount.)

The petitioners called for implementing the recommendations of the Punjab Farmers Commission and framing a policy on suicides and related issues.

In response to the government's vague affidavit, the Division Bench of Chief Justice T.S. Thakur and Justice Jasbir Singh, chastised the Parkash Singh Badal government and demanded to know what action the government had taken with regard to the clear recommendations of the Commission.

Taking up the plea of seeking directions for compensation and rehabilitation of families of farmers who had committed suicide due to financial hardships, the Bench directed the state government to file an Action Taken Report on the steps taken regarding the recommendations of the Commission.

Here is a point-by-point summary of the Punjab Farmers Commission's recommendations, followed by the point-by-point responses of the Action Taken Report submitted to the Punjab and Haryana High Court. Finally a point-by-point assessment by MASR of the government's reply has been provided.

Farmers Commission Recommendations

Relief Package for Suicide Victims' Families

1. To mitigate the sense of economic insecurity, monetary assistance should be provided on priority basis.
2. Dependents and aged women and children should be provided with monetary relief in the form of pension.
3. Psychiatric assistance should be arranged free of cost for the suicide victims' families.
4. Punjab should also be included in the category of suicide-affected states by the central government.

Preventive Measures

Streamlining credit provisions: The data collected in the study indicate that in a majority of cases, short-term credit obtained for agricultural purposes contribute to an increase in the farmers' woes. Therefore, there is a need for providing mechanisms for meeting social consumption requirements of the farmers.

Farmers' Provident Fund Scheme

A farmers' Provident Fund Scheme should be intimated to meet their social consumption needs. The farmer's contribution should be matched by the contribution by the government.

Registration of Moneylenders

The commission agents who are involved in the business of moneylending must be registered under the Moneylenders Act. There should be transparency in the accounting system and even their accounts should be computerised. The farmers should be given the statement of each and every transaction.

Payment of Farmers' Products through Account-payee Cheque

All the government agencies purchasing the produce of the farmers should make payment of the produce directly to the seller (farmer) instead of to the middle man (commission agent). With a view to ensure the payment of the produce to the farmer purchased by private trader, rule 24(ii) of the Punjab Agricultural Produce Market Act (General Rules, 1962) should be amended to make it mandatory for the payment to be made through account-payee cheque. This would lead to farmers' empowerment, empowerment by choice and empowerment of decision making.

Flood Protection

It has been observed that the farmers living on the banks of the Ghaggar river are economically most distressed, and therefore prone to suicides. The vagaries of Ghaggar in the district of Patiala, Sangrur and Mansa result in financial depredation of the farmers.

Health Insurance for the Farmers and Landless Labourers

The state should provide health insurance for the farmers and landless labourers because the uncertain health expenditure is also a major cause of debt among rural households.

Compensation Programme

A special compensation programme for crop failure, crop damage and crop diversification should be drawn up.

Checking Drug Abuse and Alcoholism

Alcohol and drug abuse, particularly among the youth, is on the rise in Punjab, especially in the suicide-prone districts. The community has to be made aware of the material and human cost of alcoholism and drug abuse borne by the society, the family and the individual, and the extent to which it diverts resources from education and human resource development. Human resource development needs to be the main thrust which would undermine the culture of habitual drinking. Drug de-addiction centres should be set up at the block level in a phased manner.

Literacy and Vocational Education Plan

The study reveals that the area which is most affected by suicides has a low literacy rate. Interestingly, the suicide victims included a large number of illiterates. Therefore, there is a need for accelerating the literacy campaign, accompanied by skill-based education.

Compensation and Rehabilitation Measures

For immediate relief to the distressed families, the Commission recommends that every farming family of suicide victim (between 20 and 65 years of age) in the state who committed suicide because of economic distress resulting from failure of crop/agricultural/allied activities or due to heavy debt, should be paid a comprehensive financial assistance as per details given below:

1. ₹50,000 should be given after the reported death with a view to tide over the immediate financial distress of the family because of the death of the breadwinner of the family.

   A pension of ₹1,500 per month to the dependents should be paid. The system of payment as followed in the old-age pension scheme should be adopted to release the monthly pension. This will be subject to the conditions whereby families in where there has been a suicide in the past 10 years, i.e., with effect from 1 January 1997, and who have reported in the police records as having a family member who has committed suicide, should be paid the above said compensation on submission of application by the families. In such cases, no further verification shall be required. The Director General of Police, Punjab, shall forward such list to the concerned Sub Divisional Magistrates (SDMs) for sanction of pension.

   The compensation of ₹50,000 should be handed over to the families of suicide victims as detailed below:
   (a) To the wife of the suicide victim.
   (b) To the parents of the suicide victim if the person was not married.
   (c) The monthly pension should be paid to the wife of the victim and to the parents if he was not married. It should

be paid to the children, if the wife remarries outside the family or is not alive, up till the age of 20 years.

Action Taken Report Submitted by Punjab Government

Relief Package for Suicide Victims

It is submitted that wide variations have been reported in the number of suicides committed by farmers on account of indebtedness by various NGOs and the district administrations. Accordingly, it has been decided that the survey of farmers' suicides on account of indebtedness be conducted by Punjab Agricultural University (PAU). Initially, they had been asked to take up two districts of Sangrur and Bathinda. Surveys in these two districts were scheduled to be completed by January 2009. Based on the experience of these two districts, survey for the other districts will be taken up thereafter. Decisions regarding the quantum of financial assistance to the affected families and the amount of pension, etc., shall be taken on completion of the survey. However, it has been decided in principle that, financial assistance of ₹2, lakh per family be given to the families of farmers and farm labourers who committed suicide on account of indebtedness since January 2000. In this respect, details of the victims and modalities to be observed are being prepared.

As far as inclusion in the category of suicide-affected states is concerned, it is submitted that this step would have serious social implications. Only the abovementioned survey would help to decide whether such a step is warranted.

Preventive Measures

The Government of India is providing short-term agricultural loans through Punjab State Cooperative Bank, District Central Cooperative Banks and Primary Agricultural Cooperative Societies, and long-term agricultural loans through Punjab State Cooperative Agricultural Development Bank and Primary Cooperative Agricultural Development Bank in the state in various schemes.

The Revolving Cash Credit Limit up to ₹6 lakh in addition to the short-term and long-term loans, is provided to the farmers to meet their other requirements. A total of 1,46,243 farmers have been sanctioned the limit of ₹2,011.13 crore.

The farmers have been given an opportunity for clearing their overdue loans by offering them a one-time settlement scheme. Under this scheme, 48,818 farmers have been provided a relief of ₹83.84 crore. Mega Lok Adalats were held on 20–21 December 2008 for the settlement of old loans of farmers and other members of the Cooperative Credit Institutions. Here as many as 46,990 old loan cases of farmers and others have been settled.

Farmers' Provident Fund Scheme

This step would have serious financial implications, and would be difficult to administer also. Hence it is not proposed to start a Provident Fund Scheme at this stage.

Registration of Moneylenders

The Chief Minister of Punjab has ordered for the creation of a law with regard to loans and interest thereon, which are advanced to the farmers and the farm labourers. For this purpose, a committee has been constituted as under:

- Chief Secretary: Chairman of committee
- Financial Commissioner Revenue: Member of committee
- Financial Commissioner Development: Member of committee
- Principal Secretary Finance: Member of committee

A preliminary meeting of this committee had been held on 26 March 2009. The matter is under active consideration of the government.

Payments to Farmers through Account-payee Cheques

At present, the seller has the option to get the payment either in cash or through cheque. But with effect from 1 July 2009, all payments shall be made to the seller through account-payee cheques only. In this respect, Notification No 11/08.06–M–3/2424, dated 16 April 2008 has been issued by the Agriculture Department (Mandi Branch).

Flood Protection

The Government of India regularly executes the work of reha-
bilitating the *bandh*s and plugging the breaches of river Ghaggar.
A project called 'Construction of Embankments and Widening of
River Ghagar from Khanouri to Village Kadail of district Sangrur'
has been sanctioned by the National Bank for Agriculture and Rural
Development (NABARD) authorities. This project aims at increasing
the capacity of the state in order to save agricultural land and village
*abadi*s from floods in river Ghaggar.

Funds to the tune of ₹3,944.46 lakhs have been released by the
NABARD authority and ₹4,132.84 lakh has been released by the state
for land payment. The work of construction of embankment from
Khanouri to Makroar Sahib (Phase I) is under progress and is likely
to be completed in current financial year, and work from Makroar
Sahib to Khanouri (Phase II) will be taken up after the completion of
work on Phase I.

Health insurance

The Bhai Ghanaya Trust has started the Bhai Ghanaya Health Care
Scheme with effect from 1 October 2008. Under this scheme,
beneficiaries can avail treatment up to ₹2 lakh (cashless) on
family-floater basis whenever required, to be admitted for a period
exceeding 24 hours for treatment. 1,83,589 lakh families have been
covered under the scheme. The treatment is available in 238 approved
hospitals and the government hospitals.

Special Compensation for Crop Failure/Damage/Diversification

The state government grants relief for damage to crops and loss
of life due to natural calamities like flood, frost, fire, etc., out of
the Calamity Relief Fund, sanctioned by the Government of India
with contribution from the state also. The rates of relief allowed by
the state government are higher than the norms sanctioned by the
Government of India. During the year 2007–08, a total amount of
₹1.15 crore was sanctioned, out of which more than ₹85 crore was
sanctioned from the state budget. Again in the year 2008–09, an

amount of ₹230 crore has been sanctioned, out of which more than ₹78 crore is from the state budget.

Checking Drug Abuse and Alcoholism

The Punjab government has undertaken a number of steps to create awareness among the youth of Punjab regarding the consequences of alcohol and drug abuse in the whole state, especially in the suicide-prone districts:

1. A Suicide Prevention Programme is being run in the department. Under this Programme, various categories of health functionaries have been sensitised about the medical aspects of suicide, and it carries awareness among the masses by holding individual meetings, group meetings and interpersonal contacts.
2. Forty-four medical officers of the Health Department have been given a three-week long training in Basic Mental Health Skill and Suicide Prevention at the Institute of Mental Health, Amritsar.
3. Twenty psychiatrists of the Health Department have been imparted training as master trainers in drug abuse prevention at the National Drug Dependence Treatment Centre, All India Institute of Medical Sciences (AIIMS), Delhi. They will impart further training to other paramedical health staff in drug abuse prevention. There are at present, eight Drug Dependence Treatment Centres (DDTC) with the Health Department. These DDTCs are established at civil hospitals in Bathinda, Jalandhar, Patiala, Faridkot, Hoshiarpur, Tarn Taran, Muktsar and Pathankot.
4. The psychiatrists at the Health Department are providing mental health care services, including treatment to drug dependent persons in district and sub-divisional hospitals in 17 districts of the state.

Twenty-two Drug Deaddiction cum Rehabilitation Centres have been established in various districts of the state by the Department of Social Security, Women and Child Development.

Literacy and Vocational Education

The recommendations of the Punjab Farmers Commission have been and are being acted upon by the government in earnest. The literacy rate for the state as per Census 2001 is 69.70 per cent. This figure should have improved to a significant level by now. The report of the National Sample Survey Organisation is awaited for the purpose.

Further, constant efforts are being made to improve literacy and to provide skill-based education to school-going children and illiterates in the age group 15–35, under the various schemes of the government, i.e., Programme of Total Literacy, Post Literacy and Continuing Education, and Sarv Shiksha Abhiyan.

A special programme of community awareness is carried out every year to apprise illiterate people about various schemes and programmes of the state and central governments, so that they can avail of the benefits.

A total budget of ₹1,447.72 crore has been approved under Sarv Shiksha Abhiyan since 2002 to 2008. An amount of ₹953.25 crore was released, of which an amount of ₹758.66 crore has been spent to date.

The state government has registered a society, namely Punjab Edusat Society, to make use of the latest technology in the education sector. The society is chaired by the Chief Minister of Punjab whereas the Executive Committee is chaired by the Chief Secretary of Punjab. The Punjab Edusat Society has already installed some 300 Satellite Interactive Terminals (SIT) in various schools, colleges, polytechnics and medical colleges who have the best available resources for the students in far-flung areas. Five hundred Receive Only Terminals (ROT) are also going to be set up in the financial year in various senior secondary schools. In addition, the schools with science stream will also be provided SITs during the current financial year itself. Further, the Department of Employment Generation is delivering a lecture every Saturday on this network, apprising the students about the latest employment opportunities available in the market.

It is felt that the efforts as indicated above will certainly help the people and particularly the illiterates and farmers in building a positive attitude towards life.

Compensation and Rehabilitation Measures

It is submitted that wide variations have been reported by different agencies on the number of suicides committed by farmers on account of indebtedness. The Police Department reported only nine cases of farmers' suicides on account of indebtedness from 2002 to 2006. As per reports received from the deputy commissioners, 132 farmers committed suicide due to indebtedness and economic hardship from the period April 2002 to March 2007.

Some NGOs also conducted studies on rural suicides. MASR, an NGO, of which Inderjit Singh Jaijee is the Convenor—on the basis of a survey of 93 villages in Moonak subdivision of district Sangrur—reported that 1,445 suicides were committed in the 93 villages of the subdivision during the period 1988–2006. Working out at the rate of 15.5 suicides per village, Jaijee estimated the number of suicides in the entire state at a minimum of 40,000, with an observation that an accurate survey may show a higher toll up to 60,000 cases during the period 1988–2006. This report seems to cover all the deceased farmers who had taken loan, and not exclusively those who committed suicide under pressure of indebtedness.

In 2006, Punjab State Farmers Commission entrusted a study to the Institute of Development and Communication (IDC), Chandigarh, on suicides in rural Punjab. IDC surveyed 600 households in 24 villages, covering areas with high, medium and low proneness to suicides. In addition, 200 families of suicide victims were interviewed. As per the report of IDC, the rate of suicide in rural population increased from 12.38 per lakh (total 1,993) in 2001 to 13.10 per lakh (total 2,292) in 2005. This report gives an estimate of suicides for the entire rural population, and not only farmers taking their lives due to indebtedness.

In view of huge variations in estimates furnished by the police, the deputy commissioners and the NGOs, it has been decided that a survey on farmers' suicide on account of indebtedness be conducted by PAU. Initially, a survey in the two districts of Sangrur and Bathinda has been completed. Based on the experience of these two district surveys, surveys of other districts will be taken up thereafter. Decision regarding the quantum of financial and other forms of assistance to the affected families and the amount of pension, etc., shall be taken on completion of the survey. However, it has been decided in principle that, financial assistance of ₹200,000 per family be given to the families of farmers and farm labourers who committed suicide on account of indebtedness since January 2000. In this respect, details of the victims and modalities to be observed are being prepared.

MASR's Assessment of the Government's Reply

The following refer to the comments by MASR on the action taken by the government on various issues raised by the Farmers Commission.

Relief Package for Suicide Victims' Families

The government-sponsored pilot census of suicides (districts of Sangrur and Bathinda) have been conducted by PAU researchers. It records 2,890 suicides. Compensation to these verified cases should be dispersed immediately. The next of kin in other districts should be granted relief as soon as the census is completed in their districts.

The Punjab government proposes to provide financial assistance to the next of kin after the entire state has been surveyed. This means that an entire generation of orphans will come of age before the government comes to their aid. Setting such a proviso is a deliberate cruelty.

The study sponsored by the Punjab Farmers Commission made this recommendation as far back as 2006 but the Punjab government has been silent on this. Punjab's farmers have been deprived of central relief package which has been extended to the states of Maharashtra, Andhra Pradesh, Kerala and Karnataka.

The Punjab government has already sponsored two studies: *Suicides in Rural Punjab* (1998) and *Suicides in Rural Punjab* (2006).[1]

Conducting yet another study through PAU is welcome in view of it being the first census survey. The study covers suicides from the period 2005 to 2008—which is against the recommendation of the

---

[1]*Suicides in Rural Punjab*. (Institute of Development and Communication: Punjab Department of Cooperation, 1998); *Suicides in Rural Punjab*. (Institute of Development and Communication, 2006).

Punjab Farmers Commission. This should cover the period from 1997 to 2008. The phenomenon of increased rural suicides began to show up in 1988 and is continuing.

Immediate relief to the suicide victim's family has been granted by the states of Maharashtra, Andhra Pradesh, Kerala and Karnataka. This recommendation of economic insecurity relief is in the right spirit. The Punjab government has faltered on this end and delayed the relief unduly.

Relief to Widows and Orphans

The government has announced ₹2 lakh for families (dependents, aged women and children). The Punjab Farmers Commission recommended ₹50 thousand as an initial grant, with ₹1,500 pension per month. This should be accepted lest the creditors grab the money. This recommendation would allow available funds to be stretched to larger number of cases. In case the government sticks to ₹2, lakh for each family, then it contradicts the announcement made in 2001 when it declared ₹2.5 lakh to the next of kin.

The real money value between 2001 and 2009 would increase further due to inflation. Hence, the government should fix the amount of compensation based on the increase in money value (taking ₹2.5 lakh in 2001) as the base.

In 2001, the then government announced a compensation of ₹2.5 lakh to next of kin. Inflation has risen, so quantum should reflect real money value.

Psychiatric Assistance

This is a good recommendation and the Punjab government should immediately implement this. It must be stressed that economic compulsion rather than insanity compelled the farmers to commit suicide.

Include Punjab in the List of Suicide-affected States

The Punjab government has expressed an aversion to claiming 'suicide-affected' status for the state. No grounds for this policy have been given. By this, the state admits that it favours deliberate concealment of rural suicides. Why should this not be deemed as a contempt of court? The following government reports have not been submitted to the Punjab and Haryana High Courts:

1.   Moonak SDM (Revenue) Report of 2001
2.   Criminal Investigation Department (CID) report of 2006
3.   IDC Report of 2006, indicating 2,000 suicides per year

The Punjab Farmers Commission estimated the suicide rate for 2001 and 2005 based on the sample survey of rural population. As per their estimate, the suicide rate was estimated at 12.38 per cent in 2001 and 13.10 per cent in 2005. The total rural population of Punjab in 2001 was 160.96 lakh. If we project the approximate number of suicides during 2001 and 2005 based on the Farmers Commission estimate, then the following numbers are obtained, as given in Table 8.1

The Punjab Farmers Commission in its report, has strongly recommended that Punjab should be accepted as a suicide-affected state like Andhra Pradesh, Karnataka, Maharasthra, etc.

The Government of India has accepted the states of Andhra Pradesh, Karnataka and Maharasthra as suicide-affected states for the purpose of central government relief, in spite of the fact that the number of suicide cases in these states is lesser than those in Punjab.

The Punjab government's deliberate refusal to declare Punjab as a suicide-affected state amounts to the violation of the citizen's Right to Life enshrined in the Indian Constitution. The Punjab government should be honest enough to accept its responsibility for farmers placed in such a gruesome plight, and should come forward like Andhra Pradesh, Karnataka and Maharasthra to accept it as a suicide-affected state.

Preventive Measures

Under negative profitability, farmers are compelled to depend on short-term credit for meeting their social consumption require-ments and survival needs. In the Action Taken Report, the Punjab government has stated that 48,818 farmers would be provided one-time settlement for clearing overdue loans. This figure represents only about 2.5 per cent of the total farming community—landowners and labourers. It is an attempt to escape responsibility. Due to the economic depression characterised by the melt-down effect, farmers and agricultural labourers are placed in a predicament. The government should advance credit to the farmers and agricultural labourers at zero per cent.

Farmer's Provident Fund

Considering the level of farmers' immiseration, this recommendation of the Farmers Commission is a necessary step. The government's refusal to implement this on grounds of 'financial implications' does not carry conviction. The government should be asked to reconsider its objection.

Registrations of Moneylenders

Punjab Relief of Agricultural Indebtedness Bill is under consideration of the state government for the past four years. It is well drafted and should be passed without further delay.

Payment to Farmers through Account-payee Cheques

This is a good recommendation and the government should implement it expeditiously.

Flood Protection

The Ghaggar Bund Project should have been undertaken in a single stage, as by building *bandh*s along the upper region only, the lower region is exposed to increased severity of flooding. A higher flood-compensation formula should be worked out in the event of future floods. The work on the lower region should be completed on priority basis.

Health Insurance

This is a very important recommendation and the government should expedite its implementation.

Crop Insurance

Crop-compensation payments should reflect the quantum of loss and not be pegged at some arbitrary figure. Farmers should be fully covered for crop loss through insurance and relief. However, presently less than 20 per cent loss is covered.

Checking Drug Abuse and Alcoholism

Excise tax on liquor sales is the state's most lucrative source of income. The state promotes liquor sales and during the present regime, villages have seen a proliferation of liquor vends. Under Prime Minister Morarji Desai, states that cut back on liquor vends were compensated for the revenue loss by the Centre. The liquor vends should be withdrawn in a phased manner and hard liquor vends should be converted into beer pubs.

Literacy and Vocational Education

The percentage of rural students in the universities of Punjab is barely 4 per cent, and in the agricultural university it is below 2 per cent. The steps to introduce vocational education are welcome but the basic remedial measures have to be taken to strengthen the quality of education at the primary and secondary level, in order to reduce the incidence of suicides. The budget allocation should be reviewed in favour of primary and secondary education.

Compensation and Rehabilitation Measures

The age of eligibility for compensation should be adjusted from 18 to 75 years. At 18, a person attains majority with full legal rights and family responsibilities. Increased longevity means that people remain active workers at 75. The compensation criteria should apply equally to men and women. The period for submitting the application for compensation should be extended to at least one year.

The Farmers Commission Report (2000) acknowledges an annual average of 2,000 debt-related rural suicides. Figures cited in the report also reveal that cases recorded by the police are negligible.

The Farmers Commission Report urges speedy compensation to the families of suicide victims but the government insists that compensation be granted only after further inquiries. The Farmers Commission report suggests an immediate relief of ₹50 thousand to families of suicide victims but the government remains silent on the quantum of relief. The government says that after the pilot survey of Sangrur and Bathinda, other districts of the state will be surveyed and only thereafter will the government decide on relief payment. One can easily see that a long time will lapse before the government gets to the point of actually paying the compensation. In the meantime many others will be following in the footsteps of fathers and brothers.[2]

---

[2]The state government had initially indicated that it was ready to pay compensation to the next of kin of suicide victims, but only after all districts were surveyed. MASR and farm unions pointed out that the time taken for the study of two districts was six months. If the study of 20 districts were conducted with the same manpower, the time taken would be five years, by which time inflation would have eaten away much of the value of the compensation. Moreover the state government had earlier proposed a compensation ₹2.5 lakh per case (the state government had set this amount in 2001). The NGOs and farm unions wanted this amount to be adjusted as per the rate of inflation during 2000–08 and the consumer price index. They also supported the suggestion of the Farmers Commission that the government should pay ₹1,500 per annum as pension to the next of kin. The NGOs and farm unions wanted the central government to declare Punjab as a suicide-affected state and extend the package of relief and rehabilitation granted to the southern states to Punjab. To exclude Punjab is discriminatory to the next of kin of suicide victims. Rural suicides in Punjab, when adjusted to the proportion of the total population of the state, are in no way less then the number of rural suicides in southern states. However, the Punjab government plainly expressed its objection to the designation and said it would not accept it. The NGOs and farm unions saw no logic in the state government's attitude.

11

Survival Strategies

LOCAL INITIATIVES ADDRESSING IMMEDIATE NEEDS OF THE WORST SUFFERERS

Rescue and Revival Mission

With both the Centre and state governments shirking their responsibility to help the suicide-affected families, it is left to the people to organise their own support system. The Rescue and Revival project of Baba Nanak Educational Society is one such endeavour.

Coming to the aid of the needy is not only a traditional practice in the villages of Punjab, it is an article of faith for Sikhs and undertaken as an obligatory religious duty. Guru Nanak preached a simple three-point formula to guide the Sikhs in the conduct of their lives: *Kirit Karo, Wand Chakko, Nam Japo* (earn by honest work, distribute your surplus, remember God) In more prosperous times, if a breadwinner died, the extended family stepped in. If the family's resources were inadequate to support the widow and orphans, then the village as a whole pitched in to ensure at least basic survival. Today the distress of rural Punjab has come to the point where every family in the village is just barely scraping through and there is no surplus with which to help others, even if people would like to do so.

Preventing more Tragedies

Throughout the past 15 years, the number of widows and orphans has grown rapidly and not a single village in the Moonak and Lehra subdivisions has been spared from suicide. By 2004, members of MASR, who had been active in enumerating and documenting suicide cases, realised that they could no longer merely collect facts. Something had to be done to keep the families of suicide victims from starving and perhaps committing suicide themselves. Aside from sustaining the families left behind, the Rescue and Revival Mission is most concerned with helping the families find hope for a better future, thereby preventing other family members from committing suicide.

Many families have sold off their last bit of land and still remain under heavy debt. Left with no means of support after the death of earning members, sometimes other members of the family are driven to take the extreme step. Cases of multiple suicides in families are becoming increasingly common. From 1988 to 2008, the number of suicide victims in two blocks of Sangrur district, namely Lehra and Andana (as documented by MASR) is 1,774. This figure is not exhaustive, there may be more. The situation is as bad in adjoining blocks.

Starvation has become a reality in rural Punjab. In most cases, next of kin of suicide victims lack resources, skills and education; they seek work as agricultural labourers on daily wages but this is not always available. In the case of the very young, the elderly and infirm, this too is beyond them. Even for the able-bodied, earning enough to simply stay alive is difficult; earning enough to educate orphaned children and provide health care for aged parents is impossible. Women in this situation are extremely vulnerable.

The idea for the Rescue and Revival Mission came from the late Daljit Jaijee, IAS. Implementation was taken up by the Baba Nanak Educational Society. It was decided to assist the neediest of families. The Rescue and Revival Mission is a non-partisan, non-sectarian organisation, providing humanitarian assistance in rural Punjab to next of kin of suicide victims. These are families with small children left completely destitute by the death of bread winners. In some cases more than one person in the family has committed suicide. These women and children have been forgotten by the government.

Keeping Families Together

After studying the situation, the Rescue and Revival members were convinced that it was a mistake to 'go the institutional way'; sending children to orphanages and elderly parents to old age homes. This would only uproot them, thereby adding to their trauma. The Mission's policy is to help the families keep their place in village society and in their own home environment. By doing so, the village Panchayats are also encouraged to come forward and help the families. Our work is to educate the children and save mothers and daughters from exploitation. So far, 160 families have been adopted.

Keeping Children in School

Education is the road to empowerment. Only those families having children of school-going age are selected for assistance through the Mission. Assistance is withdrawn if children are pulled out of the school before completing the senior-secondary level. High-school graduates with academic aptitude and interest in higher studies are awarded full fee scholarship to pursue their graduation and masters in various subjects. For girls at high-school level, training is given in computer data entry. The Nanak Singh Trust donated 30 computers for this effort.

In some cases, children have been out of school for some time (or have never gone to school) at the time of the breadwinner's suicide. Since these children have great difficulty adjusting in school, vocational education is arranged for them.

The Baba Nanak Educational Society—the umbrella organisation under which the Mission functions—set up the Jasmer Singh Jaijee (JSJ) College in Gurney Kalan, district Sangrur. This was the first college in the subdivisions of Moonak and Lehra; it is affiliated to Punjabi University. The college has made it possible for many young people to pursue higher studies.

Monitoring

To a large extent, the administrative and monitoring work of the Rescue and Revival Mission is shouldered by the faculty of the JSJ College. Lecturers of this college serve as home visitors, dropping in on the Mission families on a regular basis to see how they are getting along and particularly to see that the children are attending school. Families receive bi-monthly cheques from the Baba Nanak Educational Society. In this way the Mission ensures that assistance money spent is well utilised. Thanks to generous donors, the Mission is also able to distribute garments to needy families.

The subsistence provided to the suicide families come entirely from donations. Charitable persons have come forward to provide each family with ₹1,200 per month (₹500 for the children's education and ₹500 for family sustenance, plus ₹200 as administrative cost— total 1,200 per month).

Special mention must be made of Manjit Hardev Singh, who was the first to come forward and adopt a family. She has continued to look after her village families for many years now. So far, one of the children of the family has already completed high school.

Annual Family Adoption

Donors adopt a family and get complete details about the family they are helping. They are encouraged to visit the village and meet the families. Some donors prefer to disburse assistance themselves; for them we function as a liaison, directing them to families in dire straits. The Mission verifies each case, with complete proofs, as a 'suicide family' with no financial resources. The suicide death of the breadwinner is verified and endorsed by the Panchayat's affidavit. To ensure consistency in aiding these needy families, the minimum time period set to adopt a family is one year.

The Mission also helps the families get the pensions due to them from the state. The government provides pension amounts of only about ₹200 per month to widows, orphans, elderly destitutes and disabled persons, but for an eligible person to actually receive this amount, it requires them to complete application formalities and follow up on their applications. Successfully completing government procedures is often beyond the capacity of the poor and uneducated villagers. Out of all those eligible for pension, it is unlikely that even 10 per cent actually receive this state assistance. The Rescue and Revival Mission volunteers help the eligible to apply for pensions and follow up so that the amounts actually reach the deserving.

INTERNATIONAL RECOGNITION

In April 2009, the Rescue and Revival Mission received a powerful affirmation that its efforts were proceeding in the right direction when Harvard University's John F. Kennedy School of Government invited the Rescue and Revival Mission Project Director, Harman Kaur Sharda, to attend its annual Bridge Builders Conference at Harvard University in Boston, USA. She was one of only 20 persons from Asia, Africa and Latin America selected and the Rescue and Revival Mission was the only Indian organisation that was chosen to participate.

The Bridge Builders Conference is organised with the objective of strengthening cross-cultural understanding and grassroots cooperation between nations. Subsequently, the Kennedy School assigned one of its scholars, Mallika Kaur, the project of studying the working of the Rescue and Revival Mssion.

Some national and international NGOs extend help to destitute children by setting up 'children's homes'. The cost per child works out to be around ₹13 to 16 hundred per child per month. With less money—₹12 hundred per month—the Rescue and Revival Mission is able to cover five persons—three children, mother and a grandparent. This becomes possible as there is a substantial saving on board and lodging expense, administrative cost, school fees and transport cost. Village education is free, transport is not required and children live in their own homes. The loss of quality in education in the well-organised 'children's home' is more than compensated by the children living in their home environment. When the targeted coverage extends to thousands of children, the Rescue and Revival Project becomes more viable.

'SUICIDE FAMILIES': PUTTING FACES ON THE STATISTICS

Although statistics are revealing, telling the stories of affected families brings home the extent of the human tragedy that continues to unfold in rural Punjab. Like our statistics, the cases presented here belong to villages in Lehra and Moonak subdivisions of Sangrur district. All the families mentioned here are receiving monthly assistance ranging from ₹1,000 to 1,500. However, one could go to any district of Punjab and find families reeling under the same circumstances of exploitation, debt, dispossession, suicide and ever worsening poverty.

On the face of it, it may seem that all these cases are similar, but close acquaintance with each family reveals that each has a somewhat different story to tell. One recalls the classic opening line of Leo Tolstoy's novel *Anna Karenina*: 'Happy families are all alike; every unhappy family is unhappy in its own way.'

Village Balran

Sohan Singh's Family

Over a period of 10 years, Gurnam Kaur and Sohan Singh of village Balran lost three sons. Aaloo Singh committed suicide in 1992, Geja Singh committed suicide in 1997 and Kala Singh committed suicide in 2003. Consulting the list of suicides for Balran, one sees that in 1992 and in 1997, the family was mentioned as 'agriculturalist'— meaning that they owned at least some land. But by 2003 when Kala Singh committed suicide, the enumerator mentioned the family as 'agricultural labour', meaning that they had no holding. The family is still under a debt of ₹7 lakh from family and friends and also owes money to an *arthiya*. Now nearing 70, the elderly couple is destitute with five children to look after. But with help from the Rescue and Relief Mission, these five girls are in school, with the eldest in Class X.

Rani Kaur

Since 1988, four members of Rani Kaur's family have committed suicide. Her husband, Darshan Singh (age 25 at the time), son of Gurdev Singh, drank pesticide and died on 11 September 2003. Her brother-in-law, Nachatar Singh, drank pesticide and died on 14 April 1993. He was 40 at the time of his death. Another brother-in-law, Nikka Singh, did the same on 7 June 1988. He was 25 years old at the time and unmarried. On 11 January 2001, Nachatar Singh's son, Krishan, drank pesticide and died; he was just 19 and unmarried. Rani Kaur is now 32 years old and left alone to care for two sons and daughter—Jagseer Singh, age 15 and in Class X, Manjeet Singh, age 9 and studying in Class III and daughter Jasveer Kaur, age 8 and also studying in Class III. The family never had more than half an acre but now they have nothing. She inherited a debt of ₹80,000—₹50,000 from family and ₹30,000 from an *arthiya*.

Village Bahmaniwala

Nazma

In 2002, Hoshiar Singh (son of Suleiman), a tailor in village Bahmaniwala, finally gave up the struggle to support a family. At the age of 28, he consumed pesticide and ended his troubles. Life had been a struggle for his wife Nazma too, but with a son and two daughters to care for she felt that suicide was not an option. In the eight years that have elapsed after her husband's death, Nazma did not succeed in keeping possession of the family's one-acre holding. A debt of ₹80,000 was pending with the local *arthiya* at the time of Hoshiar's death. Two things stand out about Nazma: first that she was born deformed with arms that are no more than stumps and second, she is one of the better-educated women in the village—she has studied up to matriculation level. She has now found work in a local school. The Rescue and Revival Mission helped her get the state pensions that were due to her—widow, destitute children and handicapped person. She survives on her wages, the state's tiny pensions and the assistance from the Mission. The children—Imran Khan, Zubeida and Zarina, aged 6, 13 and 15, respectively—are all in school.

Village Bhutal Kalan

Sukhpal Kaur

After her husband Tarsem Singh (son of Surjit Singh) committed suicide, Sukhpal Kaur also attempted to kill herself but she was saved. Now she has resigned herself to life and looking after a son, Gurbinder and a daughter, Rani Kaur, aged 15 and 13 respectively. The children are studying in classes IX and VII. The family is left with just two kanals (1 kanal equals 605 square yards or $1/8$ acre) of land, hardly 2,000 sq metres.

Village Bhutal Khurd

Sansi Singh

Over the years, the joint family of Sansi Singh had accumulated a massive debt of ₹7 lakh. They started out with a holding of 6 acres but little by little they were selling bits of it. Sansi's son, Sarvir, was keenly aware that many other indebted farmers in the area had committed suicide and it worried him. He used to say that no one should attend the *bhog* (commemoration ceremony) of anyone who committed suicide. Sarvir had a son, Binder, 23 years of age, who had recently married. In fact, his bride had yet to formally shift to her new home. Like his father, Binder was under constant stress over the family's debt. In July 2001, they sold 2 acres and in December the following year, they sold another 2 acres. With hardly any land left to farm, prospects for survival were bleak. Binder took the sale in December 2002 very hard but said nothing. On 18 December 2002, when it was his turn to water the field at night, he left the house but did not come back at dawn as was usual. The family found him dead beside the tube well along with a can of pesticide that he had drunk. They had 6 acres but are now left with only half an acre and a debt of ₹7 lakh. Binder left behind Sansi Singh and Gejo Kaur, aged 65 and 55, a brother Bika Singh (25 years) and a married 23-year-old sister.

Village Chottian

Gindo Kaur

Maghar Singh had a wife, four daughters and two sons. He started out with 5 acres and used to take additional land on rent. After his sons married, they separated with each getting 1.5 acres while their father kept 2 acres. Gulzar was the younger son and in addition to his wife, he had a 3-year-old daughter and a 7-year-old son. He too used to take land on rent in an attempt to augment the meagre income from his own small holding. He took loans from both the *arthiya* and a credit society; the total amount came to about ₹1.5 lakh. The *arthiya* began demanding the return of his loan but Gulzar was in no position to pay. As the pressure mounted, Gulzar grew more and more depressed. On 1 March 2002, he attended a marriage in the house of a neighbour. The *arthiya* was also present there. Although the family did not hear the *arthiya* make any remark to Gulzar, they believe that there must have been an exchange of words because Gulzar left the marriage, returned home, shut himself in a room and drank pesticide. His wife found him dead when she returned. After Gulzar's death, his widow remarried. She left Gulzar's children with his parents but as they are themselves old and infirm, one of the married daughters, Gindo Kaur, and the older son have shared the responsibility of looking after the parents and orphans. Gulzar's land had since been sold to pay off the loan. The daughter lives in Rangewala and the son lives in Chottian.

Village Chural Khurd

Balwinder Kaur

Ruhi Ram was the elder son and at the time of his death on 16 May 2003 was 50 years old. He had a debt of ₹2.5 lakh. He was married to Balwinder Kaur and he too had a large family. Of his six daughters—Sarbjit, Sandeep, Mohinder, Jasbir, Birpal and Amandeep—the three elder daughters were married. Ruhi Ram owed ₹1 lakh to the *arthiya* and ₹1 lakh to a credit society. About a lakh was spent on digging a tube well 70 feet deep, necessitating the installation of a pump. Ruhi

Ram took an additional ₹30–40 thousand from other persons. At one time he owned a tractor but sold it off. Farming alone did not provide enough to support the family, so he also began selling milk. This brought in only a little extra money and Ruhi Ram remained constantly tense. When his third daughter married, he sent her with a dowry of a television and fridge. The expenses incurred on this marriage pushed him over the edge. The family now jointly owned only 4 acres, of which Ruhi Ram's share was one acre. There was no way he could support his family on the returns from one acre, let alone repay his debt. On 16 May 2003, he took ₹100 from his wife and told her that he was going to meet the *arthiya*. He did not return home and when his wife went looking for him, people told her that they had seen Ruhi Ram sitting near the edge of the Bhakra Canal. They asked what he was doing and he said was only resting and would take a bath after some time. Three days later on 1 May 2003, he was found hanging from a tree in his farm.

Over a period of a decade, the family of Ruhi Ram saw their landholding dwindle. His paternal grandfather, Bagga Singh left his three sons 18 acres. After Ruhi Ram's father, Gurcharan Singh, had 6 acres of ancestral land and acquired another 1.5 acre. Gurcharan Singh had five sons and a daughter. Ruhi Ram's share of the land came to 1.5 acres and one of his paternal uncles gave his land to Ruhi Ram; so initially Ruhi Ram was better off than most small farmers of Chural Khurd. But by the time of his death, debt had whittled down his holding. He sold half an acre just before he committed suicide and he was still ₹7 lakh in debt. An amount of ₹3.5 lakh is still outstanding. He had taken approximately ₹50 thousand from a regular bank but the balance is owed to an *arthiya*. He left a widow, Balwinder Kaur (now about 60 years old) and seven children—six daughters and a son. His widow sold six kanals, that is less than an acre of land, to afford the marriage of daughters and now only one acre is left. Five of her daughters are now married; one girl, Mohinder Kaur (15 years) is studying at the vocational centre and the son, Jagsir Singh (13 years) is in school. The family has been able to hold on to one acre of land but gets less from the land than it does from selling the milk of their buffaloes.

Ruhi Ram's brother had tried to commit suicide earlier. His attempt came as a reaction to a bitter quarrel about what to do about the family's debts. Most of the money had gone into getting his daughters married. The five brothers discussed selling the family's land and repaying all loans. A bitter quarrel erupted between the wives of two of the brothers, insisting that only the person in whose name the loan was taken should sell his share of the land. This greatly upset Ruhi Ram's brother. He lost his temper, went out and grabbed a can of pesticide and drank it. The family rushed him to the hospital and he was saved. The family had to spend ₹20 thousand on his treatment. The *arthiya* refused to lend money to meet these medical expenses. Ultimately, other villagers came to their aid. After some time, the brothers sold 5 acres and repaid some of the debt.

Village Dhindsa

Dhindsa, in Lehra subdivision, is a medium-sized village of 1,604 inhabitants (290 households) with a total work force of 839, of which 471 are classified as 'Main Male Workers'. Literacy rate in Dhindsa is 47.9 per cent. Between 1988 and 2006, Dhindsa had 22 suicides out of which two were women; 16 were farmers and 6 were agricultural labourers.

Mohinder Singh

In September 1998, Mohinder Singh, 32 years of age sold two of the family's 3 acres to repay a debt of ₹1.5 lakh. The sale was bitterly opposed by Mohinder's father, Labh Singh. After a violent quarrel, Mohinder drank pesticide and died. Mohinder left behind his wife, a son aged 11, a daughter aged 14, his elderly father and a very old father-in-law who was also dependent on him. The family has one acre of land left but as Labh Singh is old and often sick and the son very young, cultivating this land is difficult. The family gets some income from the sale of milk. Rescue and Revival Mission helped the family get widow and orphan pension offered by the state.

Surjit Singh

The family of Surjit Singh (27 years), son of Sher Singh, started out with 2.5 acres—a tiny holding. To partially repay a very heavy debt of ₹5 lakh, Surjit sold off about one acre, leaving the family with 1.4 acres only. By 2000, the family's situation was crushing. Trying to cope with an impossible-to-repay loan, a tiny patch of land, low profitability and on top of everything else a very bad harvest, drove Surjit to not only drink pesticide himself but also to try to force the poison down his 10-year-old son's throat. A year earlier, Surjit's wife had died in an accident. The family now consists of Surjit's two orphans—a son Harpreet and a daughter Jassi who was 8 years old at the time of her father's death—and the children's grandfather Sher Singh. Surjit Singh's death plunged his mother into an unshakable depression and she died within a few months.

Gurnam Singh

Gurnam Singh had no land and worked as an agricultural labourer. Of his three sons—Baldev, Sukhdev and Pal—two were labourers and one was a tailor in the village. Jagir Kaur, who was nearly blind, ran the house. The family's struggle to repay their loan, weighed heavily on Jagir Kaur. She became depressed and often told her neighbours that she feared that her husband or one of her sons would commit suicide. Also her failing eyesight made her feel that she was becoming a burden on the family. She began to repeat 'It would be better if I died'. On 9 January 2003, she was alone at home and drank pesticide. She is survived by her husband and sons.

Village Dudian

Amarjit Kaur

Hakam Singh, son of Gujjar Singh, farmed 2.5 acres and took some land on rent; however he barely made ends meet for his large family of a wife and five children. He took a loan of ₹1.5 lakh to dig a tube well and install a pump. The *arthiya* wrote the loan agreement deducting ₹200 for every ₹1,000. His financial position was rapidly deteriorating; among other things, he found the cost of electricity burdensome. On 1 October 2001, he went out to his field, first had some liquor and then drank pesticide and died. His family claimed that they were surprised by what he did and had no idea that he intended to do any such thing. His eldest son is now trying to repay the outstanding loan amount of ₹60,000. He left behind a wife and four children—wife Amarjit Kaur (35), sons Nirmal Singh (20), Chamkaur Singh (14), daughters Sukhpal Kaur (18) and Paramjit Kaur (16).

Village Kal Banjara

Santi Devi

Santi Devi of Kalbanjara, had six sons; two of them, Jaswant Singh and Lal Singh, committed suicide. Lal Singh committed suicide on 20 May 2002. One son is an invalid. Lal Singh's wife remarried, leaving Santi Devi to look after her three grandchildren—one girl Jasbir Kaur, 13 years in Class VII and two sons, Jaspreet Singh and Gurpreet Singh, ages 10 and 8, and studying in Classes VI and IV, respectively. The family has 2 acres left but there is no able-bodied person to farm it. The family had borrowed ₹50 thousand from other family members and have been unable to repay—a source of conflict within the wider family. In addition they owe ₹1 lakh to a government bank, ₹80 thousand to an *arthiya* and ₹50 thousand to family and friends.

Santi's father-in-law had 18 acres and three sons. Santi's husband Diwan Singh inherited 6 acres of ancestral land. The three children of Lal singh now have 1.5 acres each and Santi has 1 acre inherited from her son Jaswant singh and another 1.5 acre inherited from her other son Lal Singh.

Village Khanauri

Rameshwari

Their 65-year-old grandmother Rameshwari is all the three children, Amarjit Kaur, Rajwinder Kaur and Avatar Singh, are left with. Their parents, Basant Singh, son of Pritam Singh, and his wife Jasvir Kaur, died after consuming pesticide on 14 July 2004. The three children—aged 15, 13 and 10—are studying in classes X, VIII and VI, respectively. Their father started out with a holding of 3 acres. Now only half an acre is left and it provides hardly any income. They inherited a debt of ₹5 lakh—₹3 lakh owed to an *arthiya* and ₹2 lakh borrowed from family and friends.

At the time of his death, Basant Singh was down to 3.5 acres from 16.5 acres. He wanted to sell another half an acre but his wife, Jasvir Kaur, adamantly opposed this. After a bitter quarrel she drank pesticide and died. Hours after the cremation of Jasvir Kaur, Basant Singh too consumed pesticide and died.

Village Moonak

Phoolpati

Ram Kumar's wife, Phoolpati, and his mother used to work as agricultural labourers and there was little to distinguish the family from any other household of agricultural labourers in the village except that Ram Kumar still owned one acre. Ram Kumar agreed to farm the land belonging to an *arthiya* in return for a share of the crop. The *arthiya* met the expenses of seed, fertiliser, pesticide and supply of water. After the harvest, the *arthiya* sold the crop and kept the entire amount, refusing to give any single paisa to Ram Kumar and instead claiming that Ram Kumar owed money to him from previous loans which amounted to ₹90 thousand. After quarrelling with the *arthiya*, Ram Kumar went home, took ₹100 from his wife and bought pesticide which he drank. Phoolpati rushed him to a doctor, spending what little she had for taxi fare and medical costs, but to no avail. He died within a day. Phoolpati now survives from what she can earn as a labourer. Ram Kumar's mother also works but has found shelter with her two other sons. Phoolpati sold half an acre to get money to pay the *arthiya* but still has half an acre.

Village Nangla

Pargat Singh

Ramchand Singh and his two sons were farming 4 acres of their own land and 8 acres taken on lease. So long as the father and sons remained together, they were making ends meet without much difficulty. They had even bought a tractor, for which they had taken a loan and a diesel pump for their tube well. To return the tractor loan and buy the pump, they took a loan from an *arthiya*. The total debt was ₹2.7 lakh. But after the sons married they decided to separate leaving the sons with 2 acres each. They split the loan liability and the farming was done on just 2 acres but that did not bring in enough money to return the loan and run the household. One of the sons, Pargat (24 years), started working in a ghee factory on a wage of ₹1,900 per month. Even then, the income did not match expenses, so they sold an acre. Pargat was becoming increasingly depressed. One evening in September 2000, he told his wife that he was going out to look at the field. Instead he headed for the railway track and threw himself under a train. After his death, his father sold the remaining acre.

Major Singh

Major Singh's father, Ranjit Singh, took ₹3 lakh on loan from an *arthiya*. The burden of repayment left the family with virtually no money to survive on. In any case, they owned only 2 acres. The *arthiya* began pressurising them to return the entire amount and would often come to the house and insult them in front of the entire neighbourhood. Major Singh often remarked about other suicide cases and would say 'why kill yourself? A man should just leave the village. He should run away'. However, on 28 August 2000, after a humiliating visit from the *arthiya*, Major Singh left the house and did not return. After 3 days they found his body lying in a field with Celphos tablets beside it. He was married but childless. He left behind parents (Ranjit Singh and Labh Kaur) and elder brother Sukhbir Singh and his wife.

Angrez Kaur

Niranjan Singh (45 years), son of Dyal Singh, had a wife and four children. With a holding of about 20 acres, he was getting by comfortably. Moreover, Niranjan Singh was an *Amritdhari*[1] and spent

---

[1] A Sikh who has made a full commitment to the commandments of Guru Gobind Singh by participating in an initiatory ritual (*amrit chakna*). He takes a solemn vow to eschew alcohol (and tobacco) and keep with him the five *Kakkars*: *Kesh* (uncut hair), *kanga* (comb), *Katch* (stitched shorts), *Kada* (steel bangle) and *Kirpan* (sword, symbolic or otherwise). Also he is to reject caste and all its discriminations and mould his life according to the Sikh scriptures, which essentially means a truthful life and an honest way of life.

nothing on alcohol or other intoxicants. He managed the family finances and shared no information with the rest of the family. His wife thought that things were going well since Niranjan bought a tractor and added two rooms to the family house and got one of his daughters married. She knew that he had taken loans from the *arthiya* but was unaware of the amount. One day in June 2001, the *arthiya* came to Niranjan's home, insulted him and demanded the full repayment within 2 days, failing which he would take possession of Niranjan's land. That was when the family discovered that Niranjan owed ₹4 lakh to banks and ₹2 lakh to the *arthiya*. The family attended a wedding the next day. Niranjan was very tense and left the marriage after some time, saying that he was going home. When the family returned they found him dead. He had consumed pesticide. His widow sold 7 acres and paid back all but ₹3.3 lakh. Then her sons separated. She is left with less than 2 acres now which she farms. She also sells milk. Niranjan was survived by his wife, four sons and two daughters, one of whom is married.

Chand Kaur

The Pritam Singh household consisted of Pritam and his wife, their three sons (one married) and two daughters (both married). Although the married son, Gurtej, and his wife had a separate kitchen, the family continued to work its 10-acre holding jointly. Gurtej tilled his share of the holding and took 5 acres on lease as well, at the rate of ₹10 thousand per acre per year. The family had incurred a debt of ₹4 lakh for the marriages of the two daughters and took another loan of ₹1 lakh for Gurtej's marriage. A run of bad weather and insect infestation spoiled their crops for 3 years in a row. As they had no source of income other than what they earned from farming, they were unable to repay the *arthiya* from whom they had borrowed the money. The *arthiya* began demanding repayment. The family sold about 7.5 acres and repaid ₹2.4 lakhs of the debt. The outstanding debt and reduced holding troubled Gurtej constantly and he began drinking. On 24 November 2001, he had a couple of drinks, then went out in the veranda and consumed Celphos tablets. The family rushed him to a doctor in Sunam but he died on the way. He left behind his aged parents, his wife, two brothers and two sisters.

The Kewal Krishan family

The family of Kewal Krishan owned less than 4 acres but leased 7 acres at a rate of ₹10 thousand per acre per year. The issue of renting land and sharing the income was a source of tension in the family. The money to lease the land was taken on loan from an *arthiya* and this man was crooked. For the agricultural year of 2001, he wrote the loan agreement for ₹25 thousand and charged them interest at 30 per cent. Then he deducted ₹3 thousand as 'advance interest payment' so the family got only ₹22 thousand. The family had two other outstanding loans—one taken for a daughter's marriage and another taken to sink a tube well and buy the pump. In the first week of May, more trouble hit when the in-laws of the married sister sent the girl back to her parental home, which was a pressure tactic to get more dowry. On 8 May, Shivji Ram asked his parents for some money and was refused. This was the last straw. He consumed pesticide and died. He was survived by his parents and his sister and an elder brother who is a school teacher.

Harmeet Kaur

Tarsem (Churu) Singh (32 years), son of Ajmer Singh, started out with 5 acres in village Kadail. The family actually lived in village Moonak. In 2001, he sold his land to pay off a debt and began farming land on lease in partnership with three other men. Tarsem meanwhile took another loan of ₹4 lakh to buy a tractor. When the crop was sold, there was a dispute and his farming partners refused to pay Tarsem his share. On 3 February 2002, he came home and drank pesticide. He left behind a wife, Harmeet Kaur, and three children—Kamaljit Kaur, aged 16, studying in Class VIII; Vikramjit Singh, aged 13, studying in Class VI and Gurwinder Singh, aged 11, studying in Class V. The wife immediately sold the tractor to clear the debt and began working as an agricultural labourer at ₹50 per day. The children who had earlier studied in a private school were taken out of school.

Afterword

RHETORIC AND RESISTANCE

The union budget speeches invariably declare the budget to be 'inclusive' but they have been anything but that. The fact is that economic development and its benefits are enjoyed by a very restricted segment of the population. As for the rest of the population:

1. Poverty is not declining—in rural areas it has become endemic and fewer and fewer people can be described as even moderately well-off.
2. Unemployment is rife.
3. Access to education is becoming more restricted.
4. Economic distress has claimed the lives of hundreds of thousands of farmers and farm labourers. The union government admits that 300,000 people have taken their own lives. That distress did not just happen; the government policies created it. These policies have driven the entire rural community, including artisans and small traders, to the wall.

THE SAME OLD ALLIANCE

The political class has only one goal: capture the vote, come to power and stay in power. The business/industrial/corporate class has only one goal: capture the market, make money and keep the money. The political parties depend on the corporate sector for funds. A person standing for election in the Lok Sabha needs between 50 and 2,000 million. The political class cannot succeed without the financial backing of the business class anymore than the business class can succeed without favourable policies set by the political class. They feed each other because they need each other. Once every 5 years, the constitutionally mandated necessity of standing for election means that the electorate must be wooed with all manner of rosy pictures and sweet promises. Candidates swear by accountability, responsiveness and transparency, but this is only a temporary drama that does not change the trade-offs that actually keep the system going. After the election, people's expectations— even the most modest expectations—are ignored. Indeed, these expectations cannot be fulfilled given the commitments existing between the political and business classes.

SKEWED PRIORITIES ARE GLARING

For the rural sector of suicide-prone states, ₹710 million was granted by the Centre. (To conceal the magnitude to the problem in the state, Punjab was given only 1 per cent of this sum.) In contrast, ₹280 billion was given to Delhi for Commonwealth Games' infrastructure. Industrial development has not been equitably promoted; in particular, border states are neglected. And when it comes to agriculture, the union government clearly wishes that the 674,723,043 persons who the census listed as agriculturalists in 2001, would just disappear and leave agriculture open for corporate take-over.

The Technical Advisory Committee on Secondary Agriculture (TACSA) Report prepared for the Planning Commission by Dr D.P.S. Verma of Ohio State University spells out what will have to be done to promote 'secondary agriculture': chiefly this means reorganising agriculture production vertically, so that large corporations can control all aspects from farming per se to retail. The report says:

> Currently the majority of India's population lives in almost 600,000 small villages and are engaged primarily in agriculture and related activities. This very large labour force in agriculture has very low per capital income and due to small land holdings agriculture is becoming

non-sustainable activity to support their livelihood. Hence a substantial portion of India's current agricultural; labour force has to move to non agriculture sectors or in to agri-based industrial sector. The challenge is to manage toe transition of a large segment—perhaps even 80 per cent of the rural population from a village-centric agricultural-based economy to a city-centric non-agricultural based economy and to do so in a reasonable period.[1]

As of now, rural citizens, with a love for their land and the culture that has sprung from the village, still have the right to vote and they are unlikely to freely support any party that openly espouses the TACSA scheme. The farmer is more likely to say 'over my dead body'! Of course, it may be that the union government replies 'that can be arranged'. In fact, this is happening through policies that abet rural suicide.

CYNICISM GROWS; THE STATE IS DISCREDITED

Increasingly people at the 'crushing point' see no point in pushing for solutions within the present state system. The stability and security of the state means nothing when the individual can find neither stability nor security for his own life. Some resist and call upon the violence of the state on their heads; some give up and do violence to themselves, making their sorry ends a disgrace to the state. Violence happens in either way. Cynicism is giving way to anger and anger is becoming widespread.

For the sake of argument, let us accept that every protestor is an enemy of the state. What will we say when the ranks of the state's opponents swell to encompass 80 per cent of the population? Should we not say that the state is the enemy of the people?

In September 2010, a Hindi film actor, Salman Khan, told an interviewer in Pakistan that the 26/11 attacks were hyped because the 'elite' were targeted. Howls of fury erupted in India over this remark and the actor had to apologise profusely. He spoke the truth.

The status of the victim obviously conditions media's reaction. Suppose there are two murder victims—one a corporate executive

---

[1] *TACSA Report*, submitted to the Planning Commission, Government of India, 15 October 2008.

and one a vegetable hawker—it is no secret that greater coverage will be devoted to the executive and this is irrespective of the victim's personal merits or his situation. If the executive is a middle-aged bachelor who has not bothered to meet his parents in 10 years and the hawker leaves a widow and three children to face starvation—the executive will still receive greater coverage.

Lots of ordinary people have to die before the media takes notice and sometimes even then their deaths remain invisible—as is the case of the farmers in Punjab.

Even when lots of ordinary people die, it does not automatically mean that someone will ask why they are dying. This too is the case of the farmers of Punjab. When the farmers are committing suicide, it is common to hear, 'Very sad. They must all be drug addicts and alcoholics'.

Deliberately thought-through government policies are taking the lives of Punjab's farmers. The crime is not suicide but murder.

THE ENGULFING UNION

On 13 December 1946, addressing the Constituent Assembly, Pandit Jawaharlal Nehru moved the Objectives Resolution. He stated:

1. This Constituent Assembly declares its firm and solemn resolve to proclaim India as an Independent Sovereign Republic and to draw up for her future governance a Constitution;
2. WHEREIN the territories that now comprise British India, the territories that now form the Indian States, and such other parts of India as are outside British India and the States as well as such other territories as are willing to be constituted into the Independent Sovereign India, shall be a Union of them all; and
3. WHEREIN the said territories, whether with their present boundaries or with such others as may be determined by the Constituent Assembly and thereafter according to the law of the Constitution, shall possess and retain the status of autonomous Units, together with residuary powers and exercise all powers and functions of government and administration, save and except such powers and functions as are vested in or assigned to the Union, or as are inherent or implied in the Union or resulting therefrom; and

4. WHEREIN all power and authority of the Soverign Independent India, its constituent parts and organs of government, are derived from the people; and

5. WHEREIN shall be guaranteed and secured to all the people of India justice, social economic and political: equality of status, of opportunity, and before the law; freedom of thought, expression, belief, faith, worship, vocation, association and action, subject to law and public morality; and

6. WHEREIN adequate safeguards shall be provided for minorities, backward and tribal areas, and depressed and other backward classes; and

7. WHEREBY shall be maintained the integrity of the territory of the Republic and its soverign rights on land, sea, and air according to justice and the law of civilized nations; and

8. this ancient land attains its rightful and honoured placed in the world and make its full and willing contribution to the promotion of world peace and the welfare of mankind.

This Resolution was unanimously adopted by the Constituent Assembly on 22 January 1947.[2]

At the time of the Partition of India, Kashmir acceded to the Indian Union on the condition that it would retain a large degree of autonomy. Other Indian states had a quasi-federal status within the Union.

Fearing that other states would soon begin to press for a more Kashmir-like status, the Congress-led central government realised that Kashmir would have to be 'whittled down'.

'Whittling down' was in fact applied to all the states. Amendment after amendment enlarged the power of the Union at the cost of the states.

The union government's resiling from the initial terms of its agreement forms the genesis of today's Kashmir problem. The main reason the Kashmir problem worsens is that the union government cannot restore anything it has taken from Kashmir—cannot grant even the least significant demand—because to do so would encourage other states to demand restoration of the powers taken from them.

---

[2]This is the underlying ideology/philosophy in the Constituent Assembly. Quoted from http://parliamentofindia.nic.in/ls/debates/facts.htm (last date of access: 30 June 2011).

ENDING IMPERIAL RULE

A government that is highly centralised, suits the interests of big business. They gain a 'single-window' service. If the states were stronger and enjoyed greater authority, doing business would involve more negotiations, more interests to satisfy, more adjustments and more costs.

From a big business point of view, the only thing that could be better than the present highly centralised Indian Republic would be an Indian absolute monarchy.

A highly centralised system might work in a country where the population is basically small with less geographical and cultural diversity—a place like Switzerland or Belgium. India is highly heterogeneous. India can be strong not by steamrolling everybody but by making all its unique constituent parts strong and benefiting from cooperating with one another.

If the aim of the Indian Republic is indeed 'inclusive development', then a return to genuine federalism as envisaged at the inception of independent India is required. When states have greater powers, the ruling elite of the states will focus on the welfare of their own state. Only by doing so will they enrich themselves and remain in power (the twin aims of every politician everywhere in the world). Greater focus on the development of the state will inevitably benefit the ordinary citizen.

The sense of 'India' can take root only when citizens enjoy a feeling of empowerment at 'home level'—that is, empowerment at the level of their own cultural and economic reality. Then those who come to power can no longer regard the country as an unguarded storehouse to plunder as they like.

If there is no sense of country, there can be no patriotism; so it is little wonder that the national centre stage is a free-for-all where the powerful make the laws and at the same time consider themselves above the law.

India will benefit when each state and the people of each state are genuine participants and can clearly see that they have a stake in what happens. This does not mean some sort of a quota system or an artificial formula that stretches each region and culture on the Procrustean bed that is known in India as 'the mainstream'.

Rather every culture, region and state needs freedom to set the priorities that make sense in the context of their own resources. Devolve more powers to the states so that each is able to directly address the needs and aspirations of its people. It is time to undo the concentration of powers at the Centre.

There are fears that devolving powers to the states is a first step toward the dissolution of the Indian Union. This need not be true at all. At present, central forces are being rushed from one trouble spot to another—one day in Punjab, another day in Bastar or Manipur and yet another day in Kashmir and after that in Andhra Pradesh, and so on. Devolving power to the states will bring peace to these areas and free up resources for both the states and the union government that can be spent on productive purposes instead of efforts to just somehow keep the lid on for one more day.

Just as Punjab should not have to look to Delhi for its basic needs, neither should Karnataka, Jharkhand, Uttaranchal, Meghalaya or any other state. Now, if the all-powerful Delhi cannot or will not act, then the person in Aizawl or Bellary or Chural Kalan must simply do without. People who took care of themselves for centuries now must pin their hope on the decisions of distant ministries that are far removed from all realities except the reality of somehow coming to power in the next election.

Each state must have the autonomy and the wherewithal to pursue economic policies that will result in prosperity for its people, and this in turn will benefit India as a nation.

The following letter to Pratibha Patil, President of India, encapsulates the distress of Punjab's farmers, the injustice they have suffered and the measures needed to turn the situation around.

To: Pratibha Patil   February 12, 2010
President of India   Chandigarh
Rashtrapati Bhavan, New Delhi
RE: Farmer Suicides in Punjab

Dear Mrs Patil,

I am writing to you in my capacity as convenor of Moverment Against State Repression, Punjab, regarding the worsening situation of Punjab's small and marginal farmers and the increase in debt-induced farmer suicides in the state. Punjab is presented as the country's 'success story' and still contributes 70 per cent of wheat and 48–50 per cent of rice to the national pool (although not itself a rice-eating state) while it has only 1.5 per cent of the arable land area of India. But the rising suicides expose a vulnerability thus far cloaked, neglected and exacerbated by the Centre.

By way of background, MASR is a Punjabi non-profit [organisation] that has since 1988 worked with farmers, media, NGOs, unions, academics and think tanks to affect change in Government policies towards rural Punjab that are leading to indebtedness and suicides.

The central government selected Punjab as the recipient of its experimental Green Revolution policies. These have had several serious repercussions for Punjab, which is more agrarian than other states. For example, pressurized to turn from one crop to two crops a year, farmers are over-utilizing their fields and natural resources, while taking high loans for each crop because costs are higher than the returns the Central Government is willing to pay. Even as the negative after-effects of the Green Revolution began to be felt, the Center exacerbated the conditions by denying alternate sources of income through lack of industrialization and introducing a quota on Punjabis in the army; illegally diverting Punjab river waters necessary to sustain the type of agriculture introduced by the Green Revolution; and declining to freeze on MSPs for grains. We are noting a rise in debt-induced suicides.

MASR has recorded 1738 suicides in 91 villages from only two sub-districts of District Sangrur, one of Punjab's 20 districts. Village-wise statistics are enclosed. The recent report by Punjab Agricultural University in fact reports 2,890 suicides in the districts of Bathinda and Sangrur in 2000–08. We conservatively estimate 50,000 suicides in Punjab over the past 2 decades after taking into account that not all districts are as severely affected. Other sources support our estimates: Bhartiya Kisan Union estimates 90,000 and Punjab Farmers Commission conservatively estimates 2,000 suicides per year. Focusing on the statistics and calling for further multiple-year

Punjab and five suicide-ridden states

State 2000–08 per 1 lakh pop	Farmer suicides rural pop (cr)	Rural population (cr)	Farmer suicides per lakh population
Maharashtra	34, 659	5.58	62
Andhra Pradesh	18,396	5.54	33
Karnataka	20,592	3.49	59
Madhya Pradesh and Chhattisgarh	25,137	6.10	41
Punjab	24,732	1.61	154

Notes: (i). Farm suicide figures for Maharashtra, Andhra Pradesh, Karnataka, Madhya Pradesh and Chhattisgarh are from P. Sainath. 2010. 'Farm Suicides: A 12-year Saga', *The Hindu* , 25 January.
(ii). Punjab figures are from PAU study (April 2009) entitled *Farmers and Agricultural Labourers' Suicides Due to Indebtedness in the Punjab State: Pilot Study in Bathinda and Sangrur Districts*, Department of Economics and Sociology, Punjab Agricultural University, Ludhiana, 2009.

surveys, as has been done in the past, can no longer be the Center's response.

The incidence of farmers' suicides is much higher in Punjab compared to the other five states that have seen a high number of farmer suicides (Maharashtra, Andhra Pradesh, Karnataka, Madhya Pradesh and Chhattisgarh). This comes out clearly from Table A.1.

The farmers and their dependents, women, children, and the elderly, are extremely vulnerable. Take the case of the brothers Dulla and Bhatti, who committed suicide 7 years apart, unable to pay back the moneylenders. Their household is led by the 65-year-old mother and consists of an old father, a widow, and three children. The widow, now about 40, was first engaged to be married to Dulla. When Dulla committed suicide she was married to Bhatti, who later also drank pesticide. Her father-in-law, who lies in the backroom is depressed and had himself twice attempted suicides, worrying about the loans the family took for a tractor. Her elder son, Arjan, dropped out of 8th standard last year. Agriculture is not attractive to this child who has lost his father and uncle (and nearly his grandfather to agrarian debt) but he also does not see much future in education due to a dearth of alternate sources of employment.

We respectfully urge immediate attention and concrete action by the Center. The Central control of key agrarian policies (e.g., MSP and river waters), severely limits the State Government from bringing about the requisite change. The Center has a duty to the people of Punjab. Also, the agrarian demands today are not much different from those in the Anandpur Sahib Resolution, which preceded the farmer unrest and decade-long militancy in the State. We agree with other commentators who see suicides as violence turned inwards that will eventually turn outwards, if underlying problems are not addressed.

Given the various long-term ramifications of the urgent problem of farmer suicides, the Central Government must institute changes to its policy swiftly, holistically, and consistently. The recommended changes divided into Immediate, Short-Term and Long-Term, must be implemented within a five-year timeframe to avoid disaster.

Immediately (1 year), the Government must provide compensation to families of farmers who have committed suicide in order to respond to their plight and thwart further suicides in these families; it must provide debt waivers as in other states; must create debt conciliation boards and carefully regulate the moneylenders without risking a credit vacuum in villages.

In the Short-term (2–3 yrs), the Center must attach crop prices to the national price index, pay Punjab's electricity bill so as to free State monies for development rather than for a consumer subsidy for rest of India; increase formal/institutional credit sources to prevent the system of abuse from informal credit systems; and create pension funds and crop insurance for farmers. Make upward revision in land ceiling as small holdings are unviable. As the same time it must be ensured that only bonafide farmers of the state benefit from this measure. The laws of Rajasthan and Himachal Pradesh allow for this proviso.

In the Long-term (3–5 yrs), the Center must finally resolve Punjab's water dispute, which lies at the heart of rural despair; increase industrialization to create alternate sources of employment; and work with State Government on social education on dowry and other consumption spending as well as education sector improvements. We must note from our extensive experience with this demographic, that dowry and other spending is sometimes a precipitating factor for suicide, but not the underlying factor. Also, the capacity of these families to spend on consumption beyond necessities is very weak. Suggestions to the contrary only blame the victims, without evidence.

The time for change is now, lest Dulla-Bhatti cases keep increasing and the agrarian plight spirals out of control. The feeling of discrimination and neglect by the Government is pervasive and the Government can no longer maintain Punjab primarily as a state that continues to feed the rest of the nation while itself suffering the bitter realities harvested from the Green Revolution. We are happy to provide any additional materials and further elaborate on the recommendations. With the citizens of 91 villages of District Sangrur, we await your timely response and action.

Yours Sincerely,

Inderjit Singh Jaijee, Convenor, MASR

Cc: All Members of Parliament

Table A.1:   Village-wise debt-related suicides from 1988 to 2008: Moonak and Lehra subdivisions, Sangrur district, Punjab

Sr.no.	Village	No. of suicides	Sr.no.	Village	No. of suicides
1	Alampur	10	13	Bhai ke Pishor	7
2	Alisher	17	14	Bhatuan	23
3	Andana	47	15	Bhoolan	61
4	Arkwas	15	16	Bhunder Bhaini	5
5	Badalgarh	10	17	Bhutal Kalan	19
6	Bhamniwala	14	18	Bhutal Khurd	23
7	Bakhora Kalan	36	19	Bishenpura Khokhar	13
8	Bakhora Khurd	9	20	Bushera	27
9	Balran	85	21	Chandu	5
10	Banga	35	22	Changaliwala	6
11	Baopur	29	23	Chatha Gobindpura	15
12	Benarsi	17			

(Continued)

(Continued)

Sr.no.	Village	No. of suicides	Sr.no.	Village	No. of suicides
24	Chottian	48	53	Karoda	31
25	Chural Kalan	17	54	Khai	16
26	Chural Khurd	17	55	Khanauri Kalan	12
27	Daska	30	56	Khanauri Khurd	6
28	Dchla	20	57	Khandebad	19
29	Dhindsa	23	58	Khokhar Kalan	22
30	Dudian	29	59	Khokhar Khurd	12
31	Fatehgarh	11			
32	Gaga	29	60	Kotra Lehal	12
33	Ganauta	16	61	Kudni	2
34	Ghamur Ghat	5	62	Ladaal	23
35	Ghorenab	18	63	Lehal Kalan	52
36	Gidrani	9	64	Lehal Khurd	29
37	Gobindgarh Jaijian	26	65	Lehra	10
			66	Mahansinghwala	18
38	Gobindpura Jawaharwala	23	67	Makorar	15
			68	Mandvi	10
39	Gobindpura Papra	4	69	Maniana	19
40	Gulahri	23	70	Munak	34
41	Gurney Kalan	9	71	Nangla	31
42	Gurney Khurd	19	72	Navagaon	16
43	Hamirgarh	31	73	Phoolad	15
44	Handa	8	74	Phuleda	31
45	Hariau	15	75	Raidhrana	33
46	Harigarh	26	76	Rajalheri	2
47	Harigarh Gehla	20	77	Ramgarh Gujran	8
48	Hotipur	8			
49	Jalur	17	78	Ramgarh Sandhuan	22
50	Kadial	3	79	Rampura Gujran	14
51	Kal Banjara	23	80	Rampura Jawaharwala	13
52	Kalia	14			

(Continued)

Sr.no.	Village	No. of suicides	Sr.no.	Village	No. of suicides
81	Rattankhera	12	87	Shadihari	25
82	Rorewala	13	88	Shahpur thri	7
83	Salemgarh	19	89	Shergarh	17
84	Sangatpura	31	90	Surjan Bhaini	5
85	Sangtiwala	20	91	Thaska	17
86	Sekhuwas	12			
	Total suicides	**1,774**			

Source: This is a research report conducted by late Aman Sidhu.

Appendices

Rural Suicide: A Quantum Jump

Letter to Punjab Chief Minister Parkash Singh Badal

November 28, 2000

To: S. Parkash Singh Badal,
Chief Minister, Punjab,
Chandigarh.

Dear S. Parkash Singh Ji,

During the British Rule, States and Provinces enjoyed much greater autonomy. States had control over their resources and the legislature identified with the population with the result that in case of crisis like the present agrarian crisis, corrective measures were taken promptly.

In the beginning of the 20th century agricultural debt problem had become acute. Moneylenders had started relieving petty farmers of their land in lieu of payment of their loans. In 1929 government appointed a Banking Inquiry Committee. This committee reported that the total volume of loan in Punjab was 135 crores. It was felt that some relief to the victims of rural debt was urgently needed and a Legislative Council Committee was set up for the purpose in 1932. A bill was later passed called 'The Punjab Relief of Indebtedness Act, 1934.' Certain amendments were made to the earlier Acts relevant to the agrarian society.

These were the Provincial Insolvency Act, 1920 and Usurious Loans Act, 1918. Another Act called the Punjab Debtors Protection Act, 1936 was later introduced. Today truncated Punjab's agricultural debt is ₹ 6000 crores. The problem is thousand times more serious. Farmers faced with eviction and sell out are going in for suicide in a big way.

In most cases the debt is usurious and illegal but is taken for sheer survival. Such debts need to be settled in a just and equitable manner. This can be done by setting up Debt Conciliation Board, vesting in it the powers that it enjoyed earlier. Through this and other measures suggested in the report, farming can again be made to yield modest profits that are necessary to sustain the 75 per cent rural population of Punjab which today faces abject poverty and dislodgement.

Our report and the summary of the relevant laws are appended.

With best regards and Fateh,
Yours Sincerely,
Inderjit Singh Jaijee, Convener
Movement Against State Repression

Rural Suicides: Quantum Jump[1]

Rural suicide reports, which where earlier taboo are now beginning to appear in the press. District Ganganagar reported 175 rural suicide deaths in one year. District Amritsar reported 300 deaths during the past two and a half years. Latest study by Association for Democratic Rights (AFDR) pointed out 79 suicides in 29 villages out of which 42 suicides were for the period 1997–2000. According to AFDR suicides by Punjab farmers have increased by 250 per cent. These are sample cases, the actual number would be much higher. Haryana has reported a staggering growth in suicide cases mostly along the Punjab border. *Punjabi Tribune* of 24 October 2000 reported 10 suicides within a month in the Lehra Block; *The Hindustan Times* of 10 November 2000, talks of seven suicides within a week in district Sangrur alone.

[1]This letter is part of an enclosure attached with the letter sent by Inderjit Singh Jaijee (Convener, MASR), Baljit Kaur Gill (Co-Convener, MASR) and Aman Sidhu (Research Scholar, Department of Sociology, Panjab University) to Parkash Singh Badal, Chief Minister, Punjab.

From June to August, 2000, 40 suicides were reported from 24 villages of Lehra and Andana blocks; a list of 36 cases is attached, four names are withheld on request of family members. Considering that more than half the villages in these blocks have not been covered and also that some families like to conceal suicides, total such deaths for these two blocks would be around 50 for this period. This works out to 25 suicides per block per quarter. Suicides are reportedly more in Lehra, Andana and Barnala blocks, farmers suicides also occur more immediately after harvesting. After making appropriate discount for these two factors it may be assumed that annual suicides in these three blocks would be about 80 each per year, i.e., 240. Districts Sangrur, Mansa and Bhatinda are also badly affected districts. 14 blocks of these districts would have 40 suicides each, totaling 560. For the remaining 112 blocks of the state the rate of suicides would be 20, making a total of 2,240. The annual rural suicides in Punjab would be more than 3,000 in the rural sector.

It is strange that this level of rural suicides should remain uninvestigated by the state government and unreported by its intelligence agencies. We would be happy if media takes the initiative to investigate the 36 suicides listed by the under signed and would be more happy if we are proved wrong. The fact is that suicides are simply not reported in Punjab out of fear of police harassment and in certain cases due to fear of social stigma which sticks to the progeny of the deceased. Out of a desire to escape accountability, the government also does not like to inquire.

The government's response to these suicides has been disappointing. In 1998, government denied there were any suicides. In 1999, it admitted suicides but refused to accept that they were largely due to debt and impoverisation. In 2000, it admitted both these factors but failed to provide any direct support to the aggrieved families. The attempt of the government seems to be to bury its head in the sand and kill the problem through non-recognition.

The government had made postmortem reports essential for accepting agro-related accidental deaths including electrocution and suicides. Sensibly the government has now withdrawn this due to non-availability of postmortem facility at the block level. Instead government has made police report a must for accepting such claims. Till 1998 police records had shown zero rural suicides in Punjab. Later investigation of rural suicides ordered by the Chief Minister through police agencies brought out police obstinacy in concealing such deaths.

It is unlikely that the police would faithfully record such claims in the future. Villager's perception of the Punjab Police as a cruel and extorting agency persists. It is suggested that findings by Panchayat in this regard should be accepted as a sufficient proof.

A census on suicide should be conducted to determine the magnitude of this problem and it should not be difficult to do so, provided the government desires it, within a period of two months. The census should be conducted through the Revenue department. Medical department should also be involved in this exercise. Teams should be set up at the district level comprising a revenue officer, a doctor, members of the Kisan Unions and members of the Mazdur Union, also the members of the main political parties. This team should visit all the villages, summon the Panchayat, check the list of deaths in the village during the past 5 years and let the Panchayat confirm how many of the people died by committing suicide.

The southern states gave ₹ 1 lakh per family for repayment of loan and rehabilitation, pensions to the deserving in the family and a job along with other benefits. Punjab is given absolutely nothing on the plea that any financial help would encourage more suicides. The State Electricity Board and the Agriculture Marketing Board are already giving financial support to those who are electrocuted or die in accidents related to agricultural activities. There are no reported cases of farmers suicides in order to get these benefits. By sending an officer to visit the bereaved families to offer some financial support and on the spot pension coverage, which is in any case their entitlement, is not likely to encourage more suicides, but would help them to get over the crisis. This would also bring down the incidences of repeated suicides within the same family.

Informal rural debt was generally a confidential matter that drew a negative response from the local community. Loans by cooperative credit societies and banks gave some legitimacy to borrowing for agricultural purposes, but by and large people keep such borrowing under wraps and only reveal it when moneylenders demand either the money or the land.

With the construction of Bhakra and Sutlej–Yamuna Link (SYL) canals, natural drainage has been obstructed leading to annual floods spreading over 34 villages of Lehra and Andana blocks. No compensatory or protective measures have been taken to help out the affected villages, in spite of several court cases and assurances.

Andana and Lehra blocks receive the tail end supply of the canal system. For the past three years the canal water supply to these blocks has been reduced which is a strong contributory factor to suicides in these villages. These are critical blocks and need enhanced tail end canal water supply. Suicides in Lehra and Andana blocks discount the theory of post militancy depression as both these blocks take pride in having recorded the lowest number of killing in the state either by the militants or by the police.

Unlike farm labour, farmers receive the first impact of economic depression. Farmers have become poor relatively recently and as a result, have not yet developed the survival mechanism which farm labour has acquired over the years and therefore succumb more quickly. The farmer and the farm labourer have become equally poor. A positive side to this tragic development is a perceptible break down in the class and caste barrier through being not only co-sufferers but also co-workers. The farmers are on the fast track to becoming farm labour. At present the difference is that while the traditional farm labourer chooses to work in the same village the newly converted farm labourer, out of a sense of shame, prefers to work in the nearby villages.

Break up of joint-family system means land management is passing into less experienced hands and at the same time management cost goes up through creating two management agencies for the same farm.

Mounting farm losses are creating social tensions in the villages. On the one hand between farmers and sharecroppers and on the other hand between farmer and sharecroppers against commission agents and moneylenders. The government must find a solution to the farmers' problem before this tension reaches the stage of violence.[2]

According to data compiled by the Centre for Monitoring of Indian Economy (CMIE), Punjab grew between 1991 and 1996 at 4.6 per cent against the national average of 5.6 per cent. Today the gap is even larger.

In terms of infrastructure growth Punjab's story has been disaster-ous. It's growing at 2.1 per cent against the national average of 2.6

---

[2]Rate of growth in the agriculture sector has been only 1.7 per cent per annum, i.e., lower than the growth rate of population. Hardly one third of the Gross Domestic Product (GDP) is at present generated in agriculture and allied occupations of, the economy. But, it provides sustenance to over 70 per cent of the population. In the past 8 years, Punjab's GDP growth rate has been lower than the All-India average.

per cent. Farmers have been complaining for years about high input cost low prices of produce and diminishing profit margin. According to Punjab Agricultural University, Ludhiana, experts 'In early 1960, one quintal of wheat gave farmers enough money to buy 1000 bricks, one bag of cement and five grams of gold. However, today, one can buy just 10 bricks, a spoon of cement and one gram of gold.'

The government is trying to reduce subsidies on agriculture without controlling escalation of input prices or giving sustainable price protection. The situation could deteriorate further. Former Prime Minister, V.P. Singh has pointed out:

> United States is giving subsidies to the farmers to the extent of 9 billion dollars per year to enable them to face international competition. The latest figure given out by the Punjab Finance Minister is 32 billion dollars and 100 per cent subsidy on milk products. Indian government on the other hand is attempting to push the farmers out of production to enable foreign farmers to sell their produce in India.[3]

According to Captain Kanwaljit Singh, Finance Minister, Punjab: 'farmers are being denied direct subsidies, yet India talked of competition and quality.'[4] There is an imbalance between agriculture and industry. According to Union Minister for Agriculture Shri Sompal, 'the total volume of subsidies available to farmers is just 10 per cent, which is peanuts to what is available in the agriculturally and industrially advanced countries—60 per cent of the farmers received no subsidies in any form.' The share of agriculture in Five Year, Eight Year and Ninth Year Plans was ₹ 367 crore and 801 crore, respectively. For industry, it was ₹ 3,608 and 2,765 crore.

There has been a steady shift of population from rural to urban areas. Even today 75 per cent of the Punjab's population lives in the villages, which is sustained wholly by agricultural produce. The government imposed a land ceiling of 18 standard acres 30 years ago. Through fragmentation 80 per cent of land holdings have now come down to below 5 acres. As far back as the mid-1980s, Punjab Agricultural University, Ludhiana, had pointed out that only holdings over 14 acres were viable.

The farm reserves vanished long ago and the farmer is now trying to make two ends meet through sale of land or heavy borrowing.

---

[3] *Indian Express*, 'V.P. Singh Has a Job Cut Out for RSS', 21 October 2000.
[4] *Tribune* 'Kanwaljit warns of WTO Dangers', 5 November 2000, Chandigarh.

Here again the government has not come to the farmers aid as only 4 per cent credit comes from national banks, 15 per cent from credit societies and 81 per cent from irregular sources like *arthiyas* and moneylenders. National bank and credit societies give finance at 12 to 18 per cent interest and commission agents and moneylenders charge an interest rate varying from 24 to 40 per cent. In the latter category the principle applied is, the smaller the farmer the higher the interest which makes it more difficult for small farmers and agro labour. In contrast USA interest rate to farmers is 6 to 7 per cent.

Farming sector has been exposed to loans at 18–40 per cent rate of interest and grain price 30 per cent lower than the international price, leading to total bankruptcy of the peasantry.

The institution of *arthiyas* is essentially that of an unlicensed moneylender and also a commission agent. His earning is far more from the money lending business. It is the government's failure to provide institutional credit cover to the rural sector that has made his position so strong. Today he owns most of the village industries like brick kilns, shellers, floormills, and so on. Till a few years ago, this trade was entirely in the hands of the village bania community but now a few enterprising farmers have cut into it. What gives this trade a bad name is excessive rate of interest and its less than open accounting system. If these two aspects of the *arthiyas* can be controlled then this institution would still have a useful role to play in the rural set up. It is because of these factors that we are suggesting the setting up of a Debt Conciliation Boards for amicable settlement of rural debt. While low Minimum Support Prices (MSPs) is the most potent factor leading to rural suicides, the high rate of interest has hastened the process.

Institutional loan is only given on cash crops thereby compelling the farmer to stick to the paddy and wheat rotation. Even if the farmer wishes to diversify, storage and marketing infrastructure is not available. Through formal and informal control, the farmer is prevented from selling his produce out side the state.

Dr H.S. Shergill's study of 1998 reveals that the farmers debt in Punjab was ₹ 5,700 crore and that 70 per cent of farmers are unable to pay their loans.[5]

---

[5]H.S. Shergill, *Rural Credit and Indebtedness in Punjab.* Chandigarh: IDC, 1998. Report prepared for Punjab State Coop. Apex Bank Ltd, Department of Cooperation, Government of Punjab, pp. 88–89.

The study by Dr Gopal Iyer emphasized that indebtedness was the major causative factor for suicide. The constant pressure by lending agencies to repay the loan according to him, emerged as the most important of the precipitating factors for committing suicide.[6] Farm lands, tractors and cattle are for sale at throw away prices. Price of agriculture land all over Punjab has dropped substantially where as prices of urban properties are on the increase. If the government does not step in immediately to solve the agrarian crisis the demography of Punjab will change. Punjab farmers like those from Uttar Pradesh, Bihar will have to go to other states as farm labourers and would be replaced by the moneylenders and rich from the cities for ownership of their lands.

The government must follow the Himachal Pradesh, Jammu and Kashmir, Uttarakhand and Rajasthan example of preventing the sale of this land to non-agriculturist and people from outside the states.

They as a community bear the severest blow. Dr Parmod Kumar and Dr Gopal Iyer's research papers reveal that 72 per cent of suicides are from this community. The AFDR's report published was by Dr Sucha Singh in October 2000.[7]

The other factors that have contributed to the farmers impoverisation are diversion of three fourth of river water and hydel resources to other states in contravention of international and national laws and later on denying Punjab even the royalty on these resources and the management of its reservoirs by keeping them two to three kilometres outside Punjab during the reorganisation of state in the year 1966.

Education was taken from the State List and placed on the Concurrent List, thereby removing state control over education. States get 18 per cent for higher education and 30 per cent for school

---

[6]K. Gopal Iyer and Mehar Singh Manick, Indebtedness, Impoverishment and Suicides in Rural Punjab, Delhi: Indian Publishers Distributor, 2000, pp. 42–43.

[7]IDC Report, 1998, as cited in the Punjab Status Report 2004, submitted to the Union Government by then Chief Minister Amarinder Singh, Pramod Kumar et al. Also: AFDR Report 2000, Dr Sucha Singh, Dr Ranjit Singh Ghuman and Dr Sukhpal Singh. Also: Forced Fall, a case of Punjab Farmers, Dr T.S. Chahal, 2005, Institute of Development and Planning, Amritsar. Also Dr S.S. Johal quoted in Newsletter, Oct 15, 2006, of International Sikh Confederation. "90% of the Sikhs in Punjab reside in rural areas and their economic conditions are far from satisfactory". Dr Johl has been Economic Advisor to four Indian Prime Ministers & World Bank, also Vice Chairman, Planning Commission, Punjab, and Chairman, National Commission-Agriculture Costs & Prices.

education. Punjab has a low literacy rate.[8] This is a case of Centre denying equal access to education to rural areas and the states. The state's neglect of backward districts and mismanagement by district administration failed to promote equal educational opportunity to all the blocks. Education directly translates into employment. Rural youth, especially of the backward blocks, are denied life options and become prone to suicide.

Heavy industry had been totally denied to Punjab in order to retain it as the bread-basket of the country, without giving any compensatory advantage to the state. With the break down in agriculture, jobs in industry are just not available.

Punjab had a high representation in the army at the time of Partition and comprised of 17 per cent Sikhs and 10 per cent Hindus. After 1984, this had been reduced to 2.2 per cent, i.e., proportionate representation, bypassing the merit criteria. This formulation has also been applied to the paramilitary. Though Mr George Fernandes, the present Defence Minister has denied this, he fails to tell us the year wise percentage of recruitment of the Punjabis. It is estimated that Punjab has lost 3–4 lakh jobs. This was purely rural employment.

Agriculture is affected by vagaries of nature. For the past 50 years, crop insurance has been discussed but no concrete steps in this direction have ever been taken. Natural calamity relief is not given as a matter of right, but is disbursed occasionally as political dole; it covers only a fraction of the loss. The farmer is not pampered, least of all the Punjabi farmer.

The agricultural universities have pointed out that holdings below 14 acres are giving negative profitability. The maximum ceiling limit is 17.5 acres, only a large farmer with this kind of holding can hope to remain out of debt.

The land ceiling should be revised upward but care should be taken to ensure that Punjab's agriculture land is available only to the Punjab's agricultural community. This provision is necessary to preserve the demography of the state.

The minimum wage of a labourer is ₹ 69.05 per day, whereas according to Prof. Shergill and Parkash Singh Badal, Chief Minister,

---

[8]Sangrur district has the lowest literacy rate in the state at 45.9 per cent with Andaria, Lehra and Barnala blocks down to 28 per cent.

Punjab, an average farmer gets only ₹39 per day. Multiply this position over the years and the answer to the farmer suicide problem is revealed.

The lowest paid government servants, i.e., Class IV and III, get a starting salary of ₹4,400 and 5,200, respectively. Their last drawn salary is ₹11,421 and 12,972, respectively, in addition to pension on retirement. This is more than what a farmer with the largest permissible holding would get. This means through land ceiling laws and MSP, 75 per cent of population of India is condemned to live below or close to the poverty line. Urban and industrial sectors on the other hand are allowed to soar into the sky through earning thousands of crores. It is strange that the Congress and the Communists failed to demand and enforce urban ceiling or limit on earnings. At that time, when the Congress and Communist parties pushed through land ceilings, the Jan Sangh remained silent; fearing the axe might fall on urban property and commerce earning over a certain level as well. Leadership of all the three parties was in the hands of the urban elite.

The Green Revolution admittedly increased productivity, especially of wheat, but the benefit of this productivity was snatched from the farmers by pegging the grain price 30 per cent lower than the international price. Those who primarily profited from this Green Revolution were manufactures of farm equipment, fertilisers, insecticides and pesticides, as well as oil industries. Punjab's plea for open market price for its produce and its offer to contribute towards PDS in equal proportion to the other states was ignored. Punjab was made to bear the burden of subsiding the weaker sections of the country all these years. Industrial and urban sectors do not contribute a single penny towards PDS.

Now that the international price of the grain has become lower than domestic price, it is the duty of government to give the farmers price protection till such time as they are able to move onto viable diversification. Mr V.P. Singh, former Finance Minister and Prime Minister, has condemned the government for its 'Unilateral decision to liberalise the import of 314 agriculture items which under the WTO could have waited till 2003.'

Lately, the government's failure to purchase paddy on time, compelled the farmers to sell–off, at the least 15 lakh tonnes of paddy at a throwaway price of ₹350 to ₹400, per quintal against ₹510 and 540, per quintal. In the absence of *pucca* receipts only a few will be able to take advantage of the ₹100 crore offer made by the Centre and the state governments. Bulk of this money will find its way into the hands of corrupt officers and politicians. These farmers may not be able to pay back their loan. And, consequently, the farmer will not be able to borrow money for sowing the next crop. With the hike in diesel price, the farmers are in for harsher times and more suicides. As reported, even the Vice-Chancellor of Punjab Agricultural University, Ludhiana, Dr G.S. Kalkat, also fears a spate in suicides by the farmers and farm labourers.[9]

The Way Out: Debt Reconciliation Board

During British rule, states and provinces enjoyed much greater autonomy. States had control over their resources and the legislature identified with the population, with the result that in case of crisis like the present agrarian crisis, corrective measures were taken promptly.

In the beginning of the 20th century, agricultural debt problem had become acute. Moneylenders had started relieving petty farmers of their land in lieu of payment of their loans. In 1929, the government appointed a Banking Inquiry Committee. This committee reported that the total volume of loan in the Punjab was ₹135 crore. It was felt that some relief to the victims of rural debt was urgently needed and a Legislative Council Committee was set up for the purpose in 1932. A Bill—The Punjab Relief of Indebtedness Act, 1934—was later passed. Certain amendments were made to the earlier Acts relevant to the agrarian society. These were the Provincial Insolvency Act, 1920 and Usurious Loans Act, 1918. Another Act called the Punjab Debtors Protection Act, 1936, was later introduced.

As a consequence of these Acts, the Debt Conciliation Board was set up for amicable settlement between debtors and their creditors. Certain rules were framed concerning agricultural debt. Important rules were:[10]

1.   The court shall deem interest to be excessive if it exceeds seven and a half percent per annum simple interest or is more than

---

[9]*The Tribune*, 'Punjab Farmers Giving up Agriculture: Kalkat', 13 March 2011.
[10]The Provincial Insolvency Act, 1920; Usurious Loans Act, 1918 and the Punjab Debtors Protection Act, 1936 may be read in full at http://lawsofindia.org/statelaw.

two percent over the Bank rate, whichever is higher at the time of taking the loan, in the case of secured loans or twelve and a half percent per annum simple interest in the case of unsecured loans.

2. All the milch animals, whether in milk or in calf, kids, animals used for the purpose of transport or draught cart and open spaces or enclosures belonging to an agriculturist and required for use in case of need for tying cattle, parking carts or stacking fodder or manure, were exempted from attachment and sale in execution of a decree for realisation of debt.

3. One main residential house and other buildings attached to it (with the material and the sites thereof) and the land immediately appurtenant there to and necessary for their enjoyment belonging to a judgment debtor other than an agriculturist and occupied by him, was similarly exempted.

4. Standing crops, other than cotton and sugarcane, were not liable to attachment or sale in the execution of the decree; standing trees apart from the land on which they stand shall not be liable to sale in the execution of a decree or an order of a court.

5. Debtor's land, not exceeding 50 per centum thereof, shall not be liable to attachment or sale in the execution of a decree for the payment of money as in the opinion of the court, having regard to the judgement debtor's income from all sources except such income as is dependent on the will of another person, is sufficient to provide for the maintenance of the judgement—debtor and the members of his family who are dependent on him.

6. Agriculture covered horticulture and use of land for any purpose of husbandry inclusive of the keeping or breeding of livestock, poultry, or bees, and the growth of fruit, vegetables and the like.

7. Agriculturist included every person whether as an owner, tenant, partner, or agro labourer who depends on his livelihood mainly on income from agricultural land as defined in the Punjab Alienation Act of 1990 and every member of a tribe notified in it. In short it covered almost the entire spectrum of village community.

In addition to this, there were many other stipulations that went in favour of the farmer like exemption from arrest, etc.

The Punjab Money lending and Debtors Protection Loan (Extension and Amendment) Act, 1960, was retrograde legislation although it made a benign amendment:

> (3) such portion of the judgement debtor's land not exceeding 50 per centum thereof, shall not be liable to attachment or sale in the execution of a decree for the payment of money as in the opinion of the court having regard to the judgement debtor's income from all sources except such income as is dependent on the will of another person, is sufficient to provide for the maintenance of the judgement—debtor and the members of his family who are dependent on him.

If the debt in 1929 in united Punjab was ₹135 crore and the government of that time considered it necessary to save the peasantry from being uprooted because of the enormity of this agricultural debt, then the truncated Punjab's debt in 1998 of ₹5,700 crore is a thousand times more and needs urgent redressal. It is the paramount duty of the government today to immediately' set up a Debt Conciliation Board, vesting in it the powers that it enjoyed in 1936. This situation would not have arisen, had the power not shifted from Punjab to Delhi. The denial of the promised 'glow of freedom' has turned into a twilight foreboding dark despair for rural Punjab.

Further Suggestions

Category I

Suggestions requiring immediate action to avoid and control the increasing suicide situation:

1. **Relief to families of suicide victims**: Give ₹2 lakh to the suicide victim's family for debt relief and rehabilitation. This amount should be recovered from the Central government as the suicides are due to low MSP for the past many years.
   (a) *Declare a moratorium on interest on the existing loans*: Provide legal guidance and assistance for debt resolution in case a loan has been taken at exorbitant rate of interest

from a moneylender or other non-banking source. Set up block and village level committees across party lines, comprising of Kisan and Mazdur Unions for this purpose.

(b) Give on-the-spot pension coverage under the old age, widow, disabled and destitute categories.

(c) Panchayat confirmation of a suicide death should be accepted. Conduct census on rural suicides.

2. **Insurance:** Establish crop insurance to cover the entire farm sector.

  1. Insurance and calamity relief must cover Pest-damage and losses caused by floods, droughts and storms.

  2. Calamity relief should be given at the value of three-fourth of the value of the lost crop.

Category II
Short term measures to improve the situation:

1. Improve facilities to farmers in various sectors
   (a) Reduce interest on agricultural loans to 7 per cent.
   (b) Increase subsidy amounts and give the subsidies directly to farmers.
   (c) Make institutional loans available for cash crops to encourage diversification.
   (d) Diversification.
   (e) Encourage dehydration, cold storage, and warehousing and silo units through loans on easy terms.
   (f) Declare severely suicide-affected blocks as distressed blocks and provide additional funds for institutional support to bring them at par with other blocks.

2. Education
   (a) Appointment of more teachers in the schools and ensure proper teacher attendance. According to Dr Balkar Singh's survey on Andana block, half the schools do not have headmasters and a number of schools with not less than 300 students, have only a single teacher.
   (b) As an inducement, backward and border area allowance should be given for employees posted in such areas.

Provide increased vocational education in the rural areas.

3. Reservations

(a) The present criteria of reservation goes against the grain of our Constitution and interests of the farming community, since only persons belonging to the Scheduled Castes and Backward Classes get reservation.

(b) To correct this situation, it is necessary to base reservations on economic statuses and not caste lines. Jats must be included in the backward class category as has been done in UP or give reservation to the urban community proportionate to their population and make the rest of the jobs open to all. Even a Brahmin in the rural Punjab set up is a Dalit. This is established by the rural suicide pattern in the state.

4. Water and Irrigation

1. Take up river waters dispute in the, Supreme Court to solve Punjab's irrigation and hydel problem to bring about regular and cheaper supply to the farm sector.

2. Enhance canal water supply to tail end villages to bring the quantum supply at, par with supply to other canal irrigated areas.

Category III
General suggestions of a long-term nature:

1. Revise land ceiling upward to 80 acres.

2. Restrict land sale to Punjab agricultural classes.

3. Open trading points on the border with Pakistan for trading of agricultural produce, as is being done on our border with China and Burma for exchange and sale of local farm produce.

4. Reject World Trade Organization (WTO) for farm produce, till Punjab reaches a viable stage in crop diversification and food grain quality improvement.

5. The direction of agro research has been towards improving yield and not so much for quality improvement. More funds need to be urgently given to the research institutes for quality improvement for agricultural produce.

6. Link prices of agricultural produce to the price index.

7. Restore Punjab's earlier recruitment quota in the Defence and paramilitary services. Alternatively recruit strictly on merit, bearing in mind that Punjab has 551 kilometres of volatile and densely populated flat border to defend.

8. Establish an employment-generating heavy industry in Punjab on priority to make-up for the decades of neglect when the state was denied industry, ostensibly on account of danger from Pakistan, but in actuality to ensure that Punjab remained the nation's breadbasket.

9. Increase warehousing and silo capacity in the state in proportion to its contribution to the Centre. Grain can be transported within 5 days by road transport to any distresed area in India.

10. Transport of grains out of Punjab should only be by road transport for development of its transport industry to boost village employment. Earnings from both storage and transport should be utilized to create additional agro facilities for the farmers.

11. Encourage agro-processing and agro-based industry to provide profitability through value addition. This would also contribute towards rural profitability and employment.

12. Compensate farmers along the Indo–Pak border for restrictions placed on their farming and due to damage of their crops arising out of troop movements and restrictions on their working hours in the fields. Those, whose fields lie across the security belt, should be given 50 per cent crop loss per crop or a fixed annual compensation should not less than ₹5,000 per acre.

13. Reduce liquor vends in the villages.

14. Institutional credit should totally replace non-institutional credit. Ensure proper working of institutional financing bodies and make them more people friendly and less officious and demanding.

15. Give higher priority to encouragement of cooperative farming.

16. Provide price protection to traditional crops for the next 3–4 years till diversification reaches the desired level.

17. Insecticides and pesticides inspection and certification by competent authorities.
18. Increase direct subsidy to farm section including export subsidy. Fiscal autonomy is an absolute must for protecting states interest, which is synonymous with the farmers interest.

APPENDIX B

Summary of Laws for Regulation of Rural Debt Passed in the 1930s

In 1929, the Punjab government appointed a Banking Enquiry Committee. That Committee reported that the total volume of agricultural debt in the Punjab State was to be tune of ₹135 crore. Subsequent to this report, there was sharp fall in the prices of agricultural produce. This made the pressure of debt on the cultivator even heavier and the government felt that some relief to the victims of rural debt was urgently needed. In the beginning of 1932, Punjab government appointed a committee of members of Legislative Council to consider the problem and to submit proposals for its solution. The committee submitted its report which the Legislature debated. The government independently studied the steps taken by some other states for providing relief against indebtedness. It was against this backdrop that a Bill was introduced by the Punjab government which on its passage became the Punjab Relief of Indebtedness Act, 1934.

Before the passage of aforesaid legislation, certain Acts of the Punjab Government were in operation but they did not address the specific situation of the rural debtors. Consequently, through the Act of 1934, amendments were made in those Acts through which rural debtors could benefit.

Provincial Insolvency Act, 1920

The first Act amended was the Provincial Insolvency Act, 1920. In Section 10 of this Act, after the existing clause (a), the following clause was inserted:

(a) his debts amount to two hundred and fifty rupees, and he satisfies the Court that he is entitled to summary administration of his estate under section 74 of this Act; or The other amendment made in this Act pertains to section 74. Through the amendment, it was provided that in section 74, for the words 'five hundred rupees' the words 'two thousand rupees' shall be substituted.

Another Act which was amended was the Usurious Loans Act, 1918. In Section 3 of that Act, the following changes were made:

1. for the word 'and' in clause (a) of sub-section (i), the word 'or' shall be substituted;
2. for the word 'may' where it appears for the first time in sub-section (1) the word 'shall' shall be substituted;
3. for the word 'may' after the word 'namely' in sub-section (i) the word 'shall' shall be substituted;
4. to sub-section (2) the following clause shall be added, namely:

The Court shall deem interest to be excessive if it exceeds seven and a half per cent per annum simple interest or is more than two per cent over the Bank rate, whichever is higher at the time of taking the loan, in the case of secured loans or twelve and a half per cent per annum simple interest in the case of unsecured loans.

The above provisions, vide Section 6 of the Usurious Loans Act, 1918 were given retrospective effect which means those provisions were made applicable to all suits pending on the commencement of the Act.

A very salutary provision was made vide Part IV of the said Act under the caption Debt Conciliation Boards. In this part, a 'debtor' was defined as a person who owes a debt and,[11]

1. who both earns his livelihood mainly by agriculture, and is either a land-owner; or tenant of agriculture land, or a servant of a land-owner; or of a tenant of agricultural land, or
2. who earns his livelihood as a village menial paid in cash or kind for work connected with agriculture, or

---

[11]This enactment may be read in full of http://www.lawsofindia.org/statelaw/3295/ TheUsuriousLoansAct1918.html

3. whose total assets do not exceed five thousand rupees; provided that a member of a tribe, notified as agricultural under the 'Punjab Alienation of Land Act, 1900 (XIII of 1900), shall be presumed to be a debtor as defined in this section until it is proved that his income from other sources is greater than his income from agriculture.

The explanation added to the definition of 'debtor' provided that a debtor shall not lose his status as such through involuntary unemployment, etc., and by reason of the fact that he makes income by using his plough cattle for purposes of transport, nor shall the debtor lose his status as such only because he does not cultivate with his own hands.

The word 'agriculture' was defined as including horticulture and the use of land for any purpose of husbandry inclusive of the keeping or breeding of livestock, poultry, or bees and the growth of fruit, vegetables and the like.

Section 8 of the Act empowered the state government for the purpose of amicable settlement between debtors and their creditors to establish a Debt Conciliation Boards. Sections 9 to 29 of the Act contain provisions governing the filing of application by a debtor for settlement between him and his creditors. The application is to be verified and the particulars mentioned in the application are also listed in one of the sections. On receipt of the application, the procedure to be observed is also laid down in the sections. The Board was authorised to give notice to the creditor to submit statements of debts. One of the provisions in the sections provided that the Board shall make its best endeavour to induce the debtor and creditor to arrive at an amicable settlement.

Vide Section 15-A, further powers were conferred on the Board to adjudicate on genuineness or enforceability of debts. The Board was conferred all the powers of summoning and examining parties and witnesses, production of documents, etc., as are conferred on Civil Courts by the Code of Civil Procedure. If the parties before the Board came to an amicable settlement, the Board was authorised to reduce such settlement to writing in the form of an agreement and such agreement thus made was to take effect as if it has been a decree of Civil Court having jurisdiction in the area of jurisdiction of the Board.

The provisions of the Indian Registration Act, 1908, were made not applicable to the agreements referred to above requiring registration.

Section 20 laid down that where during the hearing of any application by the Board, any creditor refused to agree to an amicable settlement, the Board should issue a certificate to the debtor that the debtor made a fair offer which the creditor ought to have reasonably accepted. The effect of this certificate by the Board was that if the creditor sued the debtor in a Civil Court for the recovery of a debt in respect of which a certificate has been granted, the Court notwithstanding the provisions of any other law for the time being in force, shall not allow the plaintiff any costs in such suit or any interest on the debt after the date of aforesaid certification. Certain other benefits accruing to the debtor from such a certificate were also enshrined in some of the sections of the Act referred to above.

Through an amendment of Section 20-A, the decision of the Board was made final whether a loan or liability is a debt or not, or whether a person is a debtor or not.

Another section specifically barred the jurisdiction of Civil Courts to question the validity of any procedure or legality of any order or agreement made or certificate issued under the Act. Another Section provided that no appeal or application for revision shall lie against any order passed by the Board.

Yet another section laid down that when an application has been made to the Board by a debtor, no Civil Court shall entertain any new suit or other proceedings brought for the recovery of any debt and any suit or other proceedings pending before a Civil Court respect of such debt shall be suspended until the Board has either dismissed the application or an agreement has been arrived at. Even the execution of already passed decrees pending before a Civil Court was suspended during the proceedings before the Board.

Section 30 of the Act dealt with the doctrine of *dam-dupat*. It provided 'In any suit brought after the commencement of this Act in respect of a debt as defined in section 7, advanced before the commencement of this Act, no Court shall pass or execute a decree or give effect to an award in respect of such debt for a larger sum than twice the amount of the sum found by the Court to have been actually advanced, less any amount already received by a creditor'.

Section 33 of Part VII, an amendment was made of Section 1(3) (a) of the Redemption I of Mortgages (Punjab) Act, 1913. For the figures and word 130 acres the figures and word '50 acres' were substituted and in subsection (3) (b) of the same section for the figure, 1,000, the figure '5,000' was substituted.

Section 34 of the Act provided immunity from arrest. According to this section, no debtor shall be arrested or imprisoned in execution of a decree for money, whether passed before or after the commencement of the Act.

Section 35 of the Act enumerates amendments in Section 60 of the Code of Civil Procedure, 1908. Through these amendments, liberal amendments were made in favour of debtors regarding various items of property which were made exempt from attachment and sale for the realisation of debt. The salient amendments were as follows:

1. The onus was placed on the decree-holder creditor to prove that the house of the debtor was let out on rent or let to persons other than his father, mother and other family members listed therein;
2. All the milch animals, whether in milk or in calf, kids, animals used for the purposes of transport or draught cart and open spaces or enclosures belonging to an agriculturist and required for use in case of need for tying cattle, parking carts, or stacking fodder or manure, were exempted from attachment and sale in execution of a decree for realisation of debt.
3. One main residential house and other buildings attached to it (with the material and the sites thereof) and the land immediately appurtenant thereto and necessary for their enjoyment belonging to a judgement-debtor other than an agriculturist and occupied by him, was similarly exempted.

It was further provided that this protection afforded to the debtor notwithstanding, any other law for the time being in force, an agreement by which a debtor agrees to waive any benefit of any exemption under this section shall be void.

Another provision was made through the amendment of Civil Procedure Code that no order of attachment shall be made unless the Court was satisfied that the property sought to be attached was not exempt from attachment or sale.

Sections 37 and 38 provided penalties to be imposed on a creditor if he made a false claim of a principal sum or made use of documents containing false entries. In the case of a false claim of principal sum, the Court was empowered to disallow the whole claim with costs, while in case of making use of documents containing false entries, the creditor was made criminally liable, and pursuant to his prosecution and conviction, he could be awarded imprisonment up to 3 months or a fine of ₹1,000 or both.

The Punjab Debtors' Protection Act, 1936

Through this Act, the legislature considered it expedient for the more effective protection of debtors. It therefore, modified the existing law on certain points. Section 10 of this Act dealt with exemption of standing crops, trees and land from attachment and sale. It provided, notwithstanding anything to the contrary contained in any other enactment for the time being in force:

1. Standing crops, other than cotton and sugarcane, shall not be liable to attachment or sale in the execution of a decree;
2. Standing trees apart from the land on which they stand shall not be liable to sale in the execution of a decree or an order of a court;
3. Such portion of the judgement-debtor's land, not exceeding 50 per cent there of, shall not be liable to attachment or sale in the execution of a decree for the payment of money as in the opinion of the court, having regard to the judgement-debtors income from all sources except such income as is dependent on the will of another person, is sufficient to provide for the maintenance of the judgement-debtor and the members of his family who are dependent on him.

Section 12 of the Act shifted the burden of proof to the creditor. The section provided that notwithstanding anything to the contrary contained in any other enactment for the time being in force, the burden of proving that any consideration alleged to have been paid by the moneylender actually passed on shall be on him.

The Punjab Money Lenders and Debtors' Protection Laws (Extension and Amendment) Act, 1960, appears to be a retrograde legislation. The initial purpose of this Act was to extend certain moneylending and debtors protection laws to territories which immediately before the 1st November, 1956, were comprised in the State of Patiala and East Punjab States Union, but the legislature simultaneously made amendments also.

A Schedule was annexed to this Act which describe the earlier enactments which have been amended through the amendment of the Punjab Debtors' Protection Act, 1936. Sections 4, 5, 6, 6A, 7 and 8 of that Act were omitted. In Section 10 of that Act, a new clause was added:

> (3) Such portion of the judgment debtor's land not exceeding 50 per centum thereof, shall not be liable to attachment or sale in the execution of a decree for the payment of money as in the opinion of the Court, having regard to the judgment debtor's income from all sources except such income as is dependent on the will of another person, is sufficient to provide for maintenance of the judgment debtor and the members of his family who are dependent on him.

APPENDIX C

Letter to President K.R. Narayanan

May 26, 2002, Chandigarh

To: The President of India
Subject: Way to Save Debt-Trapped Farming Community

Dear Sri K.R. Narayanan,

The problem of rural debt grows increasingly serious and takes an ever-higher toll of suicide victims. Farmers and agricultural labourers, feeling trapped by their debts, see no way out except to take their own lives. Compare their response to debt with that of businessmen and industrialists. Many of them owe crores to the banks yet they shrug off their liabilities and continue to enjoy the best of everything as though they had not a care in the world.

Debt-related suicide figures from the two blocks of Lehra and Andana in district Sangrur, from April 01, 2001 to March 31, 2002, investigated by MASR and certified by village panchayats and verified by civil magistracy, stand at 56. Three cases are under investigation. A few more cases may have gone unreported due to fear of social stigma and some cases may have been sent directly to the government—the total figure would possibly be between 70 and 80. Debt-related suicides in the adjoining block of Sunam in District Sangrur and Budhiada block in Mansa District are also very high. Considering that Punjab has 138 blocks and suicides are reported from all parts of the state, the all-state figure could be as high as 4,000 a year.

On May 4, 2002, a heavily indebted young man, Jagmel Singh (27), of village Theri, Andana block, district Sangrur, ended his life, and on the very same day, 100 kms away in village Chathe Walla, district Bathinda, one Jagroop Singh (40) committed suicide for the same reason, namely debt. Both complained of harassment at the hands of commission agents, while Jagmel complained verbally, Jagroop told his story in his suicide note. He wrote: 'I do not know what will become of my children now. Please save them from the clutches of the commission agents who killed me.' Our plea is all about saving India's farmers, pushed into the clutches of commission agents.

The point is not to encourage the Punjab farmer to incur debt or behave irresponsibly where his liabilities are concerned, but he should also be aware that when he borrows money from commission agents and moneylenders in all probability he is being charged an illegal rate of interest that no court will uphold. According to a Punjab-government sponsored study by the Institute of Development and Communication (Chandigarh), nine out often Punjab farmers are in debt.

Punjab has successfully tackled rural debt earlier. In the early years of the 20th century, to meet the crisis of global depression with its devastating effects on rural economy of Punjab, the Punjab Government took drastic measures to alleviate farmers' financial hardship. In 1929, the government appointed a Banking Inquiry Committee; in 1932 it set up a Legislative Committee; and in 1934 it introduced me Punjab Relief of Indebtedness Act, 1934. A number of Acts were amended, all with the same objective of reducing the fanner's burden. Amendments were made to the Provincial Insolvency Act, 1920, and another to the Usurious Loan Act. The man instrumental in many of these measures was

Sir Malcolm Darling, ICS, whose book *The Punjab Peasant in Prosperity and Debt*, remains a landmark, and Sir Chhotu Ram, who crusaded for legislative reforms.

Their work led to the establishment of the Debt Conciliation Board in Punjab in 1934. The decision of the Board was made final and there was no appeal against it. Under the doctrine of *dam-dupat*, recovery could not be made for any sum larger than twice the amount advanced.[12] Protection was also extended to the farmer's house, implements and animals. Finally, the Punjab Debtor's Protection Act, 1936, guaranteed the inalienability of half the farmer's land and the security of crops, trees and other assets. Compare this with the present position where annually almost 3 per cent of land in rural Punjab is passing into the hands of moneylenders.

Debt conciliation is possible. As an experiment, Justice H.S. Bedi of the Punjab and Haryana High Court took up rural debt disputes in district Bathinda in 2001 through the Punjab Lok Adalat. The results have been encouraging. The farmer has seen that he has a legal option via the court rather than the illegal options of suicide or murder. But the Lok Adalat, with its handful of judges and no fixed venue, is too small an undertaking to effectively tackle the debt problems by itself. The Lok Adalat shows the way for a more comprehensive scheme.

We are asking the state to establish Debt Conciliation Boards, as was done in 1934 under Section 8 of the Punjab Relief of Indebtedness Act, 1934. When the first of such boards was established for the total united Punjab, rural debt stood at ₹135 crore. According to a Government of Punjab study in 1996–97, rural debt stood at ₹5,700 crore. Today, in much attenuated Punjab, debt stands at about ₹10,000 crore. The need for such a board is obviously very great. The government should conduct periodic studies to find out the quantum of debt. The last study (for fiscal year 1996–97) by the Institute of Development and Communication was carried out at the instance of the Punjab Government after the Movement Against State Repression repeatedly urged the state government to take note of debt-related suicides in the villages.

---

[12]The *dam-dupat* principle was established in 1934.

The present situation: According to various research papers, non-institutional debt in Punjab comprises between 60 to 80 per cent of the state's total rural debt of ₹10,000 crore. NABARD's Executive Director, S.B. Sharma has disclosed: 'nationally banks cover a mere 20 per cent of the cost of agricultural production'. He explains: 'Since agriculture is considered an unstable business, heavily dependent on the monsoons, commercial banks are reluctant to lend. Moreover, with liberalisation, the banks are cutting costs by restricting the number of small borrowers-in order to compete with foreign banks.'

The interest rate charged by the Punjab commission agent/moneylender varies from a 24 per cent to a 60 per cent (as of July/August, 2001). For the very needy, the interest may be in excess of 60 per cent. Conservatively, one may consider the average interest rate to be around 40 per cent—that is ₹400 on every ₹1,000 annually. For Andhra Pradesh, moneylenders' interest rates are generally quoted from 36 to 60 per cent and some sources quote even higher rates extending as high as 200 per cent. Similar is the position in Maharashtra. Even if interest falls this year or in future, the lower interest will not apply retroactively to loans taken earlier.

Leave aside the principal, the quantum of annual interest on non-institutional agricultural loans would be in the neighbour-hood of ₹3,200 crore. This figure is arrived at by calculating 40 per cent interest on an estimated annual rural non-institutional debt of ₹8,000 crore. This ₹8,000 crore is again an average estimated figure.

If this interest is brought down to a rate of even 16 per cent, the annual interest will drop to ₹1,280 crore from ₹3,200 crore, that is a saving of ₹1,920 crore. In other words, through administrative and legal initiative, interest on rural debt can be reduced to less than half the present figure.

How the Government Can Help

Institutional finance will help waive off the debt of marginal and small farmers altogether. If this is not possible, waive interest on the loans or impose a three-year moratorium on recovery of principal and waive interest off for farmers who are trapped in continuing losses and spiralling debt and are unable to repay their loans.

Provide *kisan* cards to all farmers to enable them to avail of bank loans. Start by issuing cards to farmers in identified suicide prone blocks.

Non-institutional finance should be outlawed. The government should enforce existing laws outlawing usurious debt.

It should make declaration of loan money over a certain amount by commission agents and moneylenders mandatory within a month of extending the loan. The government should establish Debt Reconciliation Boards on the pattern of the 1934 Act.

Considering the fact that money lent by non-institutional finance is mostly from unaccounted wealth, there is a strong possibility that a number of moneylenders would be compelled to deny outstanding debts and may go in for out-of-court settlement to the advantage of the trapped creditors. It is possible that about 80 per cent of all moneylenders would opt for out-of-court settlement while the balance 20 per cent would take their case to court.

When past interest charged by the moneylenders is accounted, then maybe there is no outstanding debt left to clear. In some cases, debtors may have repaid amounts in excess of the principle. In the case of one suicide victim of village Alampur, he took a loan of ₹3 lakh and now his heirs face a debt outstanding of ₹11 lakh as calculated by the moneylender. Under the law of *dam-dupat*, the creditor cannot recover an amount in excess of twice the principle.

Making declaration of loans in excess of a certain amount by commission agents and moneylenders mandatory would have the desired effect of unearthing black money and enforcing legal interest rates. For that matter, this would also be in the lender's interest since there would be no possibility of a debtor concealing the extent of his indebtedness.

Gain to the government would be conversion of black money into white, broadened tax base and improved circulation of white money in the market. This money perforce would be channelled into development purposes. This money so unearthed within rural Punjab could be as much as ₹8,000 crore on the basis of estimates of current non-institutional loaning. The same principle applied to the urban sector would also gamer an additional amount and an extension of this on all-India basis would bring in an astronomical figure. Even if a fraction of this hidden money were unearthed, it would strengthen the government's finances.

But by far the greatest gain would be to the well-being of rural families. Lessening of financial pressure on the farmer and farm labourer would mean few suicides and fewer families plunged into extreme hardship by the loss of breadwinners.

With this enormous money made available, banks would find it easy to establish branches in the rural areas. Unearthing black money as per our above proposal would mean that debts paid off by creditors would be substantially channelled into the banks and the banks would be made viable. Expanding the network of banks would also serve to weaken the practice of non-institutional finance with its great scope for abuse.

To spread awareness of what is and is not a legal debt liability, a team of lawyers and experts has prepared a small paper setting out the law with answers to questions on liability and legal remedies in case of illegal loans. We have tried to make the legal advice contained in the paper as comprehensive and accurate as possible in order to help the fanners and agro labourers, indeed the entire village community.

Leave aside enacting new laws for the benefit of rural debtors, even enforcing the existing law would help to bring about a salutary change, a change that would be in the interest not only of the fanner but of the state and the nation.

To the extent that non-institutional finance serves the purpose of instant credit for the farmer and till such time that the government is in a position to displace non-institutional finance in the rural areas, it may be regulated and encouraged by allowing these creditors to charge interest rates in excess of bank rates by up to 3 to 4 per cent.

Currently, there are two sources utilized by the creditor for recovery of outstanding loans: one is to apply pressure through influential persons of the village who are themselves indebted to the creditor; the second is to have the debtor harassed and intimidated by the police. Since the police has no right to interfere in civil disputes this is accomplished through bribe.

The farmers cannot escape from this debt-trap on their own. They need help. This help can come from (*a*) increasing awareness about their legal rights as debtors, (*b*) support from the village community where almost all segments of village society are in debt and (*c*) making common cause against illegal police pressure with the help of the village community. The government must help by educating the farmers on their legal rights as debtors. This can be done through (*a*) every form of mass media and awareness campaigns; (*b*) the Panchayats, revenue officials and other district officers and (*c*) debtors should be encouraged to seek the help of the courts if they have the slightest doubt that moneylenders are exploiting them.

Establishment of Debt Conciliation Boards would have the advantage of taking justice to the doorstep of the village. The Debt Conciliation Boards were fully empowered and were the final arbiter in all cases; appeals to higher courts were not allowed. Such Boards should again be set up. Till such time that these boards are set up, encourage the rural debtor to seek the protection of the courts, and make it known that money loaned in excess of a certain amount, if not declared within a stipulated periods, shall be an illegal loan and therefore non-recoverable. There should be no difficulty in enunciating this law under the Income Tax/Wealth Tax Act.

The government must find a solution to the farmers' problems before this tension reaches the stage of violence. Suicide is violence turned inward but as circumstances worsen, it is likely that violence will turn outward. From the mid-1990s, MASR has been warning the state government that neglect of the situation will lead to violence. The first signs of this are beginning to appear. As violence increases, government officials make a great outcry about law and order, but the essential underlying cause is economic desperation and injustice.

With regards,
Yours Sincerely
Sd/
Inderjit Singh Jaijee
(Convenor, MASR)
Justice A.S. Bains (President PHRO)
Baljit Kaur Gill (Co-convenor, MASR)
Lt Gen K.S. Gill (Advisor to PHRO and MASR)

Dr Gurmit Singh (Advocate)
Prof. H.S. Shergill
Gurdarshan Singh Grewal (Former Punjab Advocate General)
B.S. Toor (Advocate)
Aman Sidhu (Research Scholar)

APPENDIX D

Correspondence with the Reserve Bank of India (RBI)

August 16, 2002, Chandigarh

To: Mr Vimal Jalan, Governor
Reserve Bank of India
Central Office Building, 13th floor,Mumbai, 400001
Subject: Punjab rural suicides

Dear Mr Jalan,

We received a letter from Dr Deepali Pant Joshi (RPCD PLFS. No. 477/050104/2002-03 dated June 13) explaining the steps that the Reserve Bank is taking to ease the burden of debt on Punjab's small and marginal farmers. We are grateful to the Reserve Bank for the waiver of interest on loans to these agriculturalists covering loans not exceeding principle of ₹50,000 which have become NPAs as on 31.03.1998.

We would like to offer the following suggestions for your consideration:

1. ₹50,000 is in a very small amount where Punjab farmers are concerned as farmers at this level have been taking money for tractors, tube-wells, other agricultural machinery and fertilizers/pesticides, sometimes exceeding several lakhs and the banks have been sanctioning these loans. We suggest that the loan quantum eligible for waiver be raised to ₹2 lakh.

2. This RBI offer has to date received no publicity and is as yet unknown to the farmers despite circulars issued to

nationalized banks. To enable farmers to take advantage of the offer, it should receive wide publicity through the press and broadcast media. Where recovery has been made after the announcement of this scheme, the money should be returned to the farmers.

3. Because of insufficient publicity to the offer, it is suggested that the cut-off date be extended to give farmers a clear one-year period to settle their debts.

4. Likewise, the period covered by the scheme should be extended from 31.03.1998 to cover loans advanced up to 31.03.2001.

Our suggestions were made on the basis of normal conditions. With drought now gripping the entire region of North India, the position has become critical. We would once again suggest total waiver of loan or moratorium of three years on recovery, without interest, over and above the one-time scheme the RBI now offers. Fully 84 per cent of farmers have holdings below 5 acres; another 12 per cent hold more than 5 acres but this land has been parcelled out among sons and only fictionally remains in the name of the *karta*. It is about 4 per cent whose holdings extend up to 17 acres. A total waiver to farmers would entail very little additional cost. A copy of our letter to the Finance Minister, Government of India, is enclosed.

A government-sponsored study had placed the total rural debt of Punjab for the year 1996–97 at ₹5,700 crore. The latest figures provided by NABARD General Manager P. Satish (published in *The Tribune* of August 11) placed institutional agricultural debt for 2000–01 at ₹5,500 crore, of which ₹4,500 crore are crop loan and ₹1000 crore as agricultural term loan. NABARD's estimate for this year stands at ₹6,700 crore for both crop loan and agricultural term loan. We take it that this ₹6,700 crore includes the uncleared debt of previous years. If ₹6,700 crore is institutional debt and it constitutes 30 per cent of Punjab's agricultural debt, then the total debt would be ₹22,333 crore. (Many sources however estimate non-institutional debt at 80 per cent.) It simply means that farmers are unable to return the money and that explains the rising graph of rural suicides. Some respected agro-scientists have predicted a total collapse of agriculture in Punjab within the next 20 years. Even without the current drought, the signs of impending collapse of rural economy are clearly visible.

With warm regards,
Yours sincerely,
Inderjit Singh Jaijee, convenor

cc: Chief Minister, Government of Punjab

Reply from the Reserve Bank, 13 June 2002

Reserve Bank of India
Rural Planning & Credit Department
Central Office
Central Office Building, 13th Floor
Mumbal - 400 001

RPCD.PLFS.No.4f77/0501 .04/2002-03   June 13, 2002

Dear Sir

Punjab Rural Suicide
Please refer to your letter dated May 23, 2002. In this connection, we advise as under on the issues raised therein.

We feel that any across the board waiver of loan has an adverse impact on the financial system. However, keeping in view the difficulties faced by small and marginal farmers, we have come out with a special One Time Settlement Scheme for small and marginal farmers to be implemented by Public Sector Banks. the Scheme covers loan accounts with outstanding balance of up to principal amount (excluding any interest element) which have become NPAs as on 31.3.1998 but will not cover cases of fraud, malfeasance and wilful default. This we believe will give succour to small and marginal farmers. We enclose a copy of our circular issued in this regard for your information.

As regards the suggestion of providing Kisan Credit Card to all farmers to enable them to avail of bank loans it may be mentioned that as announced by the Union Finance Minister in his Budget Speech for the year 2001–02, commercial banks are required to cover all eligible agricultural farmers under the Kisan Credit Card Scheme within the next 3 years. We have already advised banks to take necessary steps in this direction.

The following suggestions on Non-Institutional Finance will require examination.

The Government should enforce existing laws, outlawing usurious debt.

It should make declaration of Loan money over a certain amount by commission agents and money lenders mandatory within a month of extending the loan.

The government should establish Debt Reconciliation Boards on the pattern of the 1934 Act.

Yours faithfully,
Dr Deepali Pant Joshi
General Manager

Reply by RBI (April 2003)

Reserve Bank of India
Department of Banking Operations and Development Centre –1
World Trade Centre cuffe Parade
Colaba, Mumbai – 400005

REF: DBOD Bi BP 1036/21.04.117/2002–03

Dear Sir

Punjab Rural Suicides

Please refer to your letter dated 16 August 2002 addressed to our Governor on the above subject.

We advise that we have since issued revised guidelines for compromise settlement of chronic Non-Performing Assets

(NPAs) of public sector banks. The guidelines cover NPAs in all sectors irrespective of nature of business, with outstanding balance of ₹10 crore and below on the cut-off date, i.e., 31 March 2000. A copy of our relative circular is enclosed for your information.

Yours faithfully,
M.A. Jain
Asstt. Gen. Manager

APPENDIX E

Punjab Relief of Agricultural Indebtedness Bill 2006

To streamline money leading system and to provide relief to agricultures, rural artisans and other persons dependent upon agriculture, and to provide for fair and expeditious disposal of their debt related disputes and for the matters connected there with or incidental thereto.

BE it enacted by the Legislature of the State of Punjab in the Fifty-seventh Year of the Republic of India as follows:

1.   (i)   This Act may be called the Punjab Relief of Agricultural Indebtedness Act, 2006.
    (ii)   It shall come into force at once.
2.   In this Act, unless the context otherwise requires.
    (a)   'Agriculture' shall include horticulture and the use of the land for any purpose of husbandry inclusive of the keeping or breeding of livestock, poultry, piggery, fishery or bees and the cultivation of agro forestry and the like;
    (b)   'Bank' means,
      (i)   A Scheduled Bank as defined in clause (e) of section 2 of the Reserve Bank of India Act, 1934; and
      (ii)   Any other financial institution notified by the State Government in the Official Gazette as bank for the purpose of this Act.
    (c)   'Board' means the Debt Determination and Settlement Board as provided under section 6 of this Act;

(d) 'Civil Court' includes,

    (i) A court exercising jurisdiction under the Provincial Small Cause Courts Act, 1887 (Central Act 5 of 1887);

    (ii) A court exercising powers under the Provincial Small Cause Courts Act, 1887 (Central Act 9 of 1887);

(e) 'Collector' shall have the same meaning as assigned to it under the Punjab Land Revenue Act, 1887 and includes any other officer appointed under this Act by the State Government by notification to exercise the powers of a Collector;

(f) 'Co-operative society' means a society registered or deemed to be registered under the Punjab Cooperative Societies Act, 1961;

(g) 'Debt' includes all liabilities of a debtor in cash or in kind secured or unsecured, payable under a decree or order of a civil court or otherwise, whether due or not, but shall not include,

    (i) A debt due to the Central Government or State Government;

    (ii) A debt due to a Corporation or Board or any other organization established by the Government of the State of Punjab or Government of India under the relevant law;

    (iii) Any debt incurred for the purposes of trade;

    (iv) Any rent due in respect of property let out to a debtor;

    (v) Any debt due to a bank or cooperative society;

    (vi) Any liability arising out of breach of trust or any tortuous liability;

    (vii) Any liability in respect of wages or remuneration due as salary or otherwise for services rendered;

    (viii) A debt barred by law of limitation;

    (ix) Land revenue of any other sum recoverable as arrears of land revenue; and

    (x) Any liability in respect of maintenance whether under a decree of civil court or otherwise.

(h) 'Debtor' means a person who owes a debt and

    (i) Who earns his livelihood mainly by agriculture, and is either a land-owner, or tenant of agricultural land, or a servant of a land-owner a servant of a land owner or of a tenant of agricultural land; or

    (ii) Who earns his livelihood as a rural artisan like carpenter, blacksmith, cobbler, potter or as a village menial paid in cash or kind for work connected with agriculture;

Explanation:

(a) the term debtor includes an agriculture notwithstanding the fact that he has joined service in the Armed Forces of the Union; provided that his family is engaged in agriculture.

(b) A debtor shall not lose his status as such because of the reason that

    (i) He earns income by using his plough cattle or tractor for purposes of transport; or

    (ii) He does not cultivate with his own hands; and

    (iii) "Creditor" means a person or institution to whom a debtor owes the debt.

3. Notwithstanding anything contained in any enactment for the time being in force or in any contract or other instrument having the force of law or otherwise,

(a) Every debt, together with any interest payable thereon, owed on the commencement of this Act by a debtor, shall be deemed to be wholly discharged, if

    (i) He had in the discharge of his debt, paid a sum exceeding or equivalent to double the amount of the principal at any time before the commencement of this Act; or

    (ii) He, in the discharge of his debt, pays after the commencement of this Act, a sum which, together with any sum already paid in the discharge of such debt, is equivalent to double the amount of the principal; and

(b) Every property pledged or mortgaged by a debtor whose debt is deemed to be discharged under clause (a), shall stand released and shall vest in him free from all encumbrances when such debt is deemed to be discharged.

4. There shall be moratorium on the repayment of the principal and interest, if any, by a debtor for a period of one year with effect from the date of commencement of this Act and no interest shall be charged from the debtor during the period of moratorium.

5. The interest payable on the debt shall be calculated at a rate not exceeding ten per cent annum. The interest payable on the debt shall be simple interest and not compound interest.

6. Debt Determination and Settlement Board (constitution and functions):

   (1) There shall be a Debt Determination and Settlement Board (hereinafter called the Board) art every Sub-Division in the District.

   (2) The Board shall consist of the following, namely:
       (a)   the Sub-Divisional Magistrate: Chairman
       (b)   the Sub-Divisional Magistrate: Member-Secretary
       (c)   one nominee of the Registrar, Cooperative Societies, Punjab, not below the rank of Assistant Registrar, Cooperative Societies: Member
       (d)   one nominee of the Director, Food and Supplies, Punjab, not below the rank of Assistant Food and Supplies Officer: Member
       (e)   one nominee of the Director, Agriculture, not below the Rank of Agriculture Officer: Member

7. The Board shall have jurisdiction to make settlement between the debtor and his creditor, if the total debt of the debtor does not exceed ₹30 lakh.

   (1) The quorum of the Board shall be three including the Chairman. In the absence of the Chairman, the Member-Secretary shall act as Chairman.

   (2) Where the Chairman and members of the Board are not unanimous on any point or issue, the opinion of the majority shall prevail.

   (3) If the Board is equally divided on an issue, the Chairman shall have a right to exercise a casting vote.

8. No act of the Board shall be deemed to be invalid by reason only of the existence of any vacancy of a member of the Board of any procedural defect in the transaction of business of the Board.

9. (1) A debtor or any of his creditors may apply to the Board within whose jurisdiction, the debtor resides or holds any land to determine and settle the debt-between the debtor and his creditor.

   (2) An application to the Board shall be made in writing and be signed by the applicant.

10. An application under this Act shall be presented to the Chairman and it shall contain the following particulars, namely:

    (a) In the case of a debtor,
       (i) The place where he resides or holds land;
       (ii) The particulars of all claims against him together with names and residences of his creditors;
       (iii) The particulars of his annual income from all sources;
       (iv) A statement that his total amount of debts does not exceed ₹30 lakhs;
       (v) A statement whether he has previously filed an application in respect of the same debt before the Board, and if so, with what result; and
       (vi) A statement of grounds of application and relief claimed along with copies of documents relied upon.

    (b) In the case of creditor,
       (i) the place where the debtor resides or holds land;
       (ii) the amount and particulars of his claim against such debtor;
       (iii) a statement of grounds of application and relief claimed along with copies of documents relied upon; and
       (iv) The details of registration number and licence number given as a money lender under the Punjab Registration of Money Lenders Act, 12938.

          Provided that no claim of creditor shall be entertained by the Board, if he is not registered as a money lender under the Punjab Registration on Money Lenders Act, 1938.

11. (1) On receipt of an application under section 9, the Board shall pass an order fixing a date and place for hearing the application.

(2) The Board may, pending its final decision on application, make such interlocutory order including the order for stay of recovery of debt, as it may deem necessary in the interest of justice.

(3) A notice of the order under sub-section (1) or sub-section (2), as the case may be, shall be sent to the creditor against who application has been submitted by the debtor by such means, as the Board may consider most effective and speedy, at the cost of the applicant and where the debtor is not the applicant, the notice of the orders referred to above, shall be sent to him in a similar manner.

(4) The Board and the Collector may, adopt the procedure for conducing the proceeding under this Act, including the service as provided in the Punjab Land Revenue Act, 1887, for the proceedings before a revenue officer.

12. (1) On the date fixed under sub-section (1) of section 11, the Board shall issue notice to the creditor of the debtor to submit a statement of debt owed to such creditor by the debtor. Such statement shall be submitted to the Board in writing giving details of every entry of debit and credit in the account of the debtor within a period of one month alongwith the copies of debtor with in a period of one month alongwith the copies of documents to be relied upon the support of his claim:

Provided that the Board may extend the period for submission of statement of debts for the reasons, to be recorded.

Provided further that the total extended period shall not exceed two months.

(2) Every debt owed to a creditor of which no statement has been submitted to the Board in compliance with the

provisions of sub-section (1), shall be deemed to be duly discharged for all purposes and all occasions against such creditor, and every debt owed to two or more creditors jointly, of which such statement or statements signed by all such creditors or their recognised agents has or have not been so submitted, shall be deemed to be so discharged against such creditors as have failed to submit the said statement but only to the extent of their respective shares in the said debt:

Provided that no such debt shall be deemed to be discharged against any creditor whose name has not been included in the application made under section 9.

(3) If the creditor or any of the joint creditors fails without sufficient cause to be present in person or by his recognised agent or, with the permission of the Board by legal practitioner, in accordance with the provisions of Section 23, at any of the hearings fixed by the Board, or fails to produce full particulars and documents as required under sub-section (1) of section 12, the debt due to him or to the joint creditors, as the case my be, shall be deemed for all purposes and all occasions to have been fully discharged.

(4) If a debtor fails to appear before the Board without sufficient cause after due service, the Board shall proceed ex-parte to determine the debt.

13. (1) Every creditor submitting in compliance of a notice issued under sub-section (1) of the section 12, a statement of the debts owed to him, shall furnish, alongwith the statement, full particulars of all such debts, and shall at the same time, produce all documents (including entries in books of account) on which he relief to support his claims, together with a copy of every such document.

(2) With a view to ascertain the originally and correctness of the documents produced under sub-section (1), the Board shall make through verification thereof and after ascertaining the original to the creditor.

14. (1) If a creditor or a debtor as the case may be, challenges the genuineness of enforceability or the total amount of debt or principal or rate of interest or amount of interest or date of raising of any debt include in an application, the Board shall adjudicate upon the issue.

(2) The Board shall determine the case of each debt shown in the application made by the debtor or his creditor under section 9 or in the statement furnished by the creditor under section 12, other than a debt, declared non-genuine or unenforceable, the principal amount originally advanced, he amount paid by the debtor towards the principal or interest or both and the amount of principal and interest payable strictly in accordance with the principles laid down under this Act.

(3) The Board shall decide every application submitted under section 9 within a period of six months from the date of application, unless the Board, for the reasons to be recorded in writing deems it fir to extend the period further upto a period of not more than six months.

(4) Where the debtor is found to have repaid to the creditor an amount equal to or exceeding double the principal amount, or the debtor or being apprised of such finding, pays an amount, which makes the total repayment equal to double the amount of principal, the Board shall declare the debt as fully discharged and thereupon the provision of clause (b) of section 3, shall apply.

(5) The Board shall keeping in view the outstanding amount of principal and interest and interest as determined under sub-section (2), and the repaying capacity of the debtor, order the repayment of the amount of debt settled or determined in half-yearly equal instalments, but not exceeding ten instalments.

(6) If the Board finds that the provisions of the Punjab Regulation of Accounts Act, 1930, have not been complied with by the creditor, the Board may disallow the interest partly or wholly for the period of non-compliance.

(7) No order passed under this section shall be called in question in a civil court.

15. The Board may exercise all such powers connected with the summoning, the examining of parties and witness, and with the production of documents, as are exercised by a Revenue Officer under the Punjab Land Revenue Act, 1887.

16. Notwithstanding anything contained in any other law for the time being in force, if the Board has reason to believe:

   (a) That the interest charged by the creditor, is excessive or

   (b) That the transaction was as between the parties thereto, substantially unfair;

   The Board shall exercise all or any of the following powers namely:

   (i) Re-open the transaction, take an account between the parties and relieve the debtor of all liabilities in respect of any excessive interest;

   (ii) Notwithstanding any agreement, purporting to close previous dealings and to create a new obligation reopen any account already taken between them and relive the debtor of all liabilities in respect of any excessive interest, and if anything has been paid or allowed in account in respect of such liabilities order the creditor to repay any sum, which it considers to be repayable in respect thereof;

   (iii) Set aside either wholly or in part or revise or alter any security given or agreement made in respect of any loan, and if the creditor has parted with the security, order him to indemnify the debtor in such manner and to such extent, as it may deem just.

   Provided that, in the exercise of these powers, the Board shall not,

   (i) Re-open any agreement purporting to close previous dealings and to create a new obligation which has been entered into by the parties or any persons from whom the twelve years from the date of the transaction;

   (ii) Do anything which affects any decree of a civil court passed before the commencement of the Act.

Explanation:

1. In the case of an application brought on a series of transactions, the expression "transaction" means for the purpose
of clause (1) of the proviso, the first of such transactions.

2. The interest shall be deemed to be excessive, if it exceeds the
rate of interest mentioned in section 5.

17. If once an application has been made by a debtor and it
has been disposed of, the Board shall not entertain second
application within a period of two years of the disposal of
the first application.

18. If any question arises in any proceedings under this Act as
to whether a loan or liability is a debt or not, or whether
a person is a debtor or not, the decision of the Board shall
be final, and shall not be called in question in any civil court.

19. No civil court shall entertain,

(a) Any suit, appeal or application for revision

(i) To question the validity of any procedure or the
legality of any order or award made under this
Act; or

(ii) To recover any debt, which is deemed to have
been duly discharged under the provisions of this
Act; or

(iii) Any suit for declaration or any suit or application
for injunction affecting any proceedings under
this Act before the Board.

20. (a) Any person, who, from the discovery of any new and
important fact or evidence which after the exercise of
due diligence, was not in his knowledge or could not
be produced by him before or at the time, when an
order was made by the Board, or on account of some
mistake or error apparent on the face of the record,
or for any other sufficient reason, desires to obtain a
review of such order, he may submit an application
to the Board for reviewing such order:

(b) On receiving an application under sub-section (1) the
Board may, if no appeal has been filed, by reviewing
its order, pass such order, as it may deem appropriate:

Provided that the Board shall not pass on order which adversely affects any interested person without giving such person an opportunity of being heard.

Provided further that no application for review shall be entertained, if presented more than sixty days after the date of passing the order, sought to be reviewed.

21. Any person aggrieved by an order of the Board under section 14, may file an appeal before the Collector within a period of 60 days of the date of receipt of the order. No appeal shall lied from an order, refusing to review or confirming on review.

22. The Financial Commissioner, Revenue may, at any time, call for the record of any case pending before, or disposed of, by any Board or Collector and may, pass such order, as he thinks appropriate:

Provided that he shall not pass an order reversing or modifying any proceedings or order of the Board or Collector, as the case may be, without giving an opportunity of being heard to the concerned parties.

Explanation:
In this section, the expression 'Financial Commissioner' shall have the same meaning as assigned to it in the Punjab Land Revenue Act, 1887. (Act No XVII of 1887)

23. In any proceedings under this Act, party may be represented by an agent, duly authorised in writing or with the permission of the Board by a legal practitioner.

24. (a) On and from the commencement of this Act, no civil court shall entertain any new suit filed for the recovery of any debt, covered under this Act.

    (b) Any suit pending before a Civil Court on such commencement in respect of any such debt, shall be transformed by the court to the Board having jurisdiction.

25. The order of the Board passed under this Act, shall be executed by the civil court having jurisdiction in the area as if it were a decree or order of that court.

26. Overriding effect: The provisions of this Act shall have an overriding effect over the provisions contrary to it in any other law for the time being in force.
27. The State Government may, by notification in the Official Gazette, make rules for carrying out the provisions of this Act.

APPENDIX F

Village Profiles

The Census of Punjab 2001 helps to understand the nature of the Moonak *tehsil*'s 91 villages. The suicides referred to occurred between 1988 and 2007. The youngest victim was 18 years old while the oldest was 70 years plus.

Lehra Block

Lehra Block comprises 47 villages that have been studied in totality along with four villages namely Fatehgarh, Ghuduan, Phuleda and Gobindgarh Jaijian which, although they fall in the Sunam *tehsil*, are attached to Lehra Police Circle. Lehra Block does not face annual floods but very rarely it gets partially flooded.

Alampur has a population of 1,355 with 225 households. It has a recorded number of 458 total workers out of which 296 are main workers, 162 marginal workers and only 897 non-workers. Alampur's literacy rate is 41.3 per cent. This village is predominantly 'Punjabi'. Village Alampur has had a spate of suicides since 1992 as compared to other villages. There were no farmer suicides found before 1992. Alampur has had 9 suicides (*including 1 case of multiple suicides*) out of which, 62 per cent are in the age group of 21 to 30 years and two from the age group of 41 to 50 years. Only one of victim was 20 years old. Of the nine, four were farmers and five were farm labourers. All the suicides were committed by consuming pesticide or poison.

Alisher has a population of 1,421 with 258 households and 463 literates, having a low literacy rate of 38 per cent. It has a total of 659 workers out of which 380 come under the category of main workers and 279 marginal workers. Alisher had 17 suicides (*including 2 cases of multiple suicides*) with the maximum number (5) in 1999. Three women of Alisher committed suicide in 1994, 1997 and 1999. Pushpa Devi (1999) has left behind three daughters aged 12, 9 and 6. Out of these 17 suicides, 7 belong to the labour class and the remaining 10 belong to the Jat community. Alisher is a Punjabi non flood-prone village. Most cases fall between the age group of 20 to 35 years, with two cases aged from 40 to 45 years group. Most of the victims committed suicide by consuming pesticide but three drowned in the nearby canal, one burned himself and one hung himself.

Arkwas with a population of 4,586, has 727 households. Its total workers comprises 50.76 per cent of the total population, with a literacy rate as high as 51.3 per cent. Arkwas has had 14 suicides out of which 5 belonged to the labour class and the remaining 9 were Jat Sikh agriculturists.

Bakhora Kalan and Bakhora Khurd Bakhora Kalan has a population of 2,284 with 393 households and a total of workforce 859, out of which 628 are main workers. Bakhora Kalan had 34 suicides (*including 6 cases of multiple suicides*) out of which 14 belonged to the agricultural community, 1 was a tailor and the remaining 19 were agricultural labourers. Out of these, one was a woman. In case of Naib Singh, aged 25, who committed suicide on 16 July 1994, his brother Nek Singh committed suicide in 1986. Naib Singh, an agriculturist, committed suicide after selling his last bit of land. Bakhora Kalan has had 11 suicides in the age group from 18 to 20.

Bakhora Khurd has a population of 1,133 with 588 total workers and 187 households. Its literacy rate is very low at 37.2 per cent. Bakhora Khurd has had nine suicides (*including 1 case of multiple suicides*) out of which five were agricultural labourers and the remaining were farmers. Most of the cases were between the age of 30 and 50 years.

Balran's population is 6,566; it has 1,111 households and 2,856 total workers, out of which there are 2,470 main and 386 marginal workers. Balran's literacy rate is 39.7 per cent. The village has had 81 suicides (*including 15 cases of multiple suicides*), which is the highest in all of Lehra Block. Out of these, four were women. A sudden increase in suicides was seen in 1995, 1996 and 1997 when 15, 13 and 9 persons took their own lives in these 3 years alone. Two suicides, that of Sukhdev Singh, son of Bachan Singh, who committed suicide on 19 August 1990, aged 32 and Sansi Singh, aged 25, who committed suicide on 13 October 1992, were brothers and farmers. Another multiple suicide in this village was of Gurdev Singh's sons and grandson; Nikha Singh, the grandson of Sukhdev Singh in 1988, Nachattar Singh, aged 40, in 1995 and Darshan Singh on 9 November 2003. They were agricultural labourers. Balran has had 48 suicides of farmers and 33 of labourers.

Bhai Ke Pishor has a population of 3,073 and 563 households, with 1,394 total workers and literacy rate of per cent. It has had four suicides; three of them were agricultural labourers and one was a farmer.

Bhatuan This village has a population of 1,891 with 312 households and 1,094 total workers with 47.7 per cent literacy rate. Bhatuan has had 22 suicides (*including 2 cases of multiple suicides*), out of which 6 were agricultural labourers and 16 were farmers. Bhatuan has had a father and son suicide. The father Kartar Singh, aged 50, committed suicide in 1993 and the son, Leelu Singh, aged 25, committed suicide in 1988. Both killed themselves by hanging.

Bhutal Kalan and Bhutal Khurd Bhutal Kalan has a population of 4,988, having 822 households and 2,000 total workers. The literacy rate here is only 38.4 per cent and the sex ratio (0–6) is only 653. Bhutal Kalan had 17 suicides out of which 2 were agricultural labourers and the rest were farmers. Tarsem Singh, aged 40 years, committed suicide on 12 November 2000 along with his wife but she was saved.

Bhutal Khurd has 453 households with a population of 2,656, out of which 902 are workers. The village has a literacy rate of 38.4 per cent. Bhutal Khurd has had 17 suicides (*including 3 cases of multiple suicides*), out of which 12 were farmers and five were labourers.

Changaliwala This village has a population of 988 with 174 households and 535 total workers. Five people committed suicide; out of which two were farmers and the remaining three were agricultural labourers.

Chottian has a population of 2,960 with 528 households. Its literacy rate is only 38 per cent and it has 1,152 main workers and 173 marginal workers. In Chottian, 46 persons committed suicide (*including 5 cases of multiple suicides*) within the years 1994 to 1997, registering the highest number. Three women have committed suicide in Chottian. In the case of Pala Singh, son of Magar Singh, aged 21, his elder brother also committed suicide a few years earlier. In Chottian, 23 suicides have been of farmers and 23 of agricultural labourers.

Chural Kalan and Chural Khurd Chural Kalan has 529 households with a population of 3,130, out of which 1,001 are workers. The literacy rate of 56.3 per cent is relatively higher than the previously mentioned villages. Sixteen persons committed suicide out of which two were farmers, 13 were agricultural labourers and one was in service.

Chural Khurd has a population of 1,173, with 177 households and 469 total workers. Twelve persons committed suicide (*including four cases of multiple suicides*); out of these three were farmers and nine were agricultural labourers.

Daska has 688 houses with 4,010 as the total population. Here the non-workers are almost double: the total work force comprises 2,635 main workers and 1,375 marginal respectively. Since 1990, Daska has had 30 persons committed suicide in this village (*including 2 cases of multiple suicides*); 16 were farmers, 11 were agricultural labourers and one was a serviceman. Of these, six were women and eight suicides were by unmarried persons.

Dhindsa has a population of 1,604 with 290 households and a total worker force of 839 and 471. Literacy rate in Dhindsa is 47.9 per cent. Dhindsa has had 23 suicides (*including 2 cases of multiple suicides*) out of which two were women; 15 were farmers and six were agricultural labourers. In the year 2000, Shera Singh, aged 27, committed suicide along with his son but the son survived.

Dehla has a population of 1,713 with 319 households and 623 main workers, which means that on an average, two persons in each family work. It has a literacy rate of only 37.9 per cent. Since 1988, 20 persons of this village have committed suicide (*including 1 case of multiple suicides*), out of which 12 were agriculturalists and six were agricultural labourers.

Gaga has a population of 2,881, with 518 households and 1,248 total workers. Twenty-seven persons committed suicide, out of which 11 were farmers and 16 were agricultural labourers. One woman committed suicide on 17 October 2003.

Ghorenab has a population of 3,242 and 577 households. Out of its 1,145 total workers, 1,115 are main workers. It has a literacy rate 38.3 per cent. Fifteen persons have committed suicide, out of which one was a Brahmin woman, 10 were farmers and 4 were agricultural labourers.

Gidrani has 256 households and a population of 2,069 populations. It has a total of 707 workers, which makes it almost 2.76 workers per household; out of these 590 are main workers. Nine persons have committed suicide; five are farmers and four agricultural labourers.

Gobindpura Papra with a population of 1,301 and 232 households, five persons have committed suicide. It has a total work force of 861, which amounts to almost 3.71 workers per family. There has been no suicide among the farmers.

Gobindpura Jawaharwala has a population of 1,663 with 272 households but only 515 total workers. Its literary rate is 47.3 per cent; 22 persons committed suicide in this village (*including 2 cases of multiple suicides*), out of which one was a woman. There are two cases of individuals who have attempted suicide repeatedly but have been saved. They are dealt with in this study as special cases as they are in a position to say what drove them to take this step. Both sons of Karnal Singh committed suicide: Darshan Singh who attempted suicide on 4 June 2004 and his younger brother, Lakha Singh, aged 32, who committed suicide on 8 January 2005. When Parkash Singh attempted suicide, the family had 2 acres but by the time his younger brother committed suicide they were reduced to 1.5 acres and had been able to reduce their debt by ₹60,000 only. Out of the total suicides, 14 were of farmers, seven of labourers and one was in service. Two persons above 60 years of age committed suicide.

Gurney Kalan and Gurney Khurd The Census of Punjab 2001 has clubbed these two villages together. According to the census, they have a total population of 1,400 with 229 households. They have a total of 521 workers and 879 non-workers. The literacy rate of these two villages is 36.8 per cent and the sex ratio is 867, whereas the sex ratio of ages 0–6 is 979. Nine persons have committed suicide in Gurney Kalan, out of which eight were farmers and one was a labourer. Nineteen persons have committed suicide in Gurney Khurd (*including 3 cases of multiple suicides*), out of which 17 were of farmers and two were labourers. In an area with a population of 1,400, suicides are many.

Hariau has a population of 2,470 and 448 households with 953 workers. Fourteen persons have committed suicide in this village (*including 1 case of multiple suicides*), out of which eight were farmers and six were agricultural labourers.

Jalur has 512 households with a total population of 2,986 and 1,381 total workers, out of which 1,317 are main workers and 64 marginal workers. Jalur has a literacy rate of 42 per cent. Sixteen persons have committed suicide (*including 3 cases of multiple suicides*), out of which 11 were farmers and five were agricultural labourers.

Kal Banjara has 290 households with a total population of 2,388. It has 711 total workers and 1,677 non-workers with 39.6 per cent literacy rate. Fourteen persons have committed suicide (*including 2 cases of multiple suicides*), out of which eight were farmers, six were agricultural labourers and one was a 19-year-old girl. In the case of suicide victim Jagdish Singh, he belonged to an agricultural family with 3 acres but had to sell all his land to repay his debt. Two brothers, Lal Singh and Jaswant Singh, sons of Diwan Singh, committed suicide. Lal Singh died on 20 May 2002 and his brother Jaswant Singh committed suicide in 2000.

Khai No data was available from the Census about Khai as it has been included in Lehra Township for data collection. Village Khai has had 16 suicides (*including 1 case of multiple suicides*), out of which four were farmers and 12 agricultural labourers. One of the victims was a woman. Preetu, wife of Pal Singh, committed suicide in 1995 and her son Nirmal, aged 18, an agricultural labourer, committed suicide on 20 September 2005.

Kalia again for census purposes has been added in Lehra Township. So its individual statistical data was not available. Kalia has had four suicides, three were agricultural labourers and one was a farmer.

Khandebad has a population of 2,118 with 354 households and 777 total workers, which is almost half of the non-workers—which is 1,404. Its literacy rate of 53.1 per cent is relatively higher than the other villages in this region. Khandebad has had 17 suicides (*including 1 case of multiple suicides*), out of which four were farmers and 13 were agricultural labourers.

Khokhar Kalan and Khokhar Khurd For the purpose of Punjab Census data collection, Khokhar Kalan and Khokhar Khurd have been clubbed together. Their population is 2,718 with a total of 462 households and 1,351 total workers, out of which 1,140 are main workers and 211 marginal workers. Its literacy rate is 44.8 per cent. In Khokhar Kalan, 22 persons had committed suicide (*including 1 case of multiple suicides*); 16 were farmers and six were agricultural labourers, including a woman. Chhaju Singh, aged 50, son of Kesar Singh, who committed suicide in 1998 and Makhan Singh, son of Chhaju Singh, aged 30, who committed suicide in 2001, were father and son. Khokhar Khurd has 12 suicides (including 1 case of multiple suicides), out of which eight were farmers and four were agricultural labourers. For a population of 1,663 and 279 households, 34 suicides makes it 2.04 per cent of the total population of the villages.

Kotra Lehal has a population of 2,257 with 391 households and 1,301 total workers. The literacy rate of the village is 47.2 per cent. Ten persons have committed suicide, out of which seven were farmers and two were agricultural labourers and one was in service.

Ladaal has a population of 2,459 and 415 households. It has 1,189 total workers out of which 540 are main workers and 649 marginal workers. The literacy rate of Ladaal is 47.6 per cent. Twenty persons have committed suicide (*including 4 cases of multiple suicides*), out of which four were of women. In all, there were 11 farmers and nine agricultural labourer families.

Lehra government has included Lehra village in Lehra Township for census purposes, so its individual data was unavailable. Ten persons have committed suicide in this village, out of which one was a woman. Lehra has had two suicides from an agricultural family, five who were agricultural labourers and three who were in service or self-employed such as shopkeepers, cycle repairmen and carpenters.

Lehal Kalan is a big village with a population of 5,308 and 917 households. It has 2,355 total workers and 2,953 non-workers. The literacy rate of the village is 39.7 per cent. Since 1993, 49 persons have committed suicide (*including 2 cases of multiple suicides*) in Lehal Kalan. It is the village with the *second highest number* of suicides after Balran in the Lehra Block. The prevalence of suicides rose suddenly in 1998 with 15 suicides, three women have committed suicide; 44 were farmers and five were agricultural labourers. There have been two cases of double suicides since 1993. Sukhchain Singh, aged 23, who committed suicide in November 2000 and Jinder Singh, aged 22, who committed suicide on 10 May 2002, were brothers. Parkash Begum, who committed suicide on 7 November 2002 and Anguri Begum, who committed suicide on 17 April 1999, were sisters married to brothers.

Lehal Khurd has a population of 2,787 with 460 households and 41.6 per cent literacy rate. Twenty-six persons have committed suicide (*including 2 cases of multiple suicides*), out of which 19 were farmers and seven were agricultural labourers.

Nangla has a population of 3,008 with 522 households and a total working population of 1,313, out of which 1,175 are main workers and 138 marginal workers. The literacy rate of Nangla is 43 per cent. Twenty-nine persons have committed suicide *(including 2 cases of multiple suicides)*, out of which 17 were farmers and 12 were agricultural labourers. One of the victims was a woman.

Raidhrana is a big village with 4,921 population and 826 households. But the total working population of 2,099 is not even half of the total population. Main workers (1,621) account for one-third of the total population. Raidhrana has had 15 suicides by farmers, seven by agricultural labourers and the rest by farmers. Since 1988, every year has brought one or two suicides in Raidhrana.

Ramgarh Sandhuan has 423 households with a total population of 2,373. The census lists 1,056 workers and gives the literacy rate as 46 per cent. Fifteen persons have committed suicide *(including 1 case of multiple suicides)*, out of which 12 were farmers; three were agricultural labourers and all under the age of 30; the eleventh victim was a woman aged 52.

Rampura Jawaharwala with a population of 1,272 and 231 households, is a small village with only 374 main workers and three marginal workers. The literacy rate is 47.3 per cent. Ten persons have committed suicide *(including 1 case of multiple suicides)*, and all of them have been of farmers.

Rattankhera is a village where persons have committed suicide *(including 1 case of multiple suicides)*. All were farmers. The suicide rate is not exceptional given the population of 1,792 and 304 households. The village literacy rate is 56.2 per cent, which is relatively higher than the other villages.

Rorewala has a population of 1,701 with 317 households and only a total working population of 679. This village has the highest percentage of workers to total population and the second highest gender ratio of 10:27 in the 0–6 age group. (In terms of gender ratio, Rajalheri in Andana block stands number one with a ratio of 1031.) Twelve persons have committed suicide *(including 1 case of multiple suicides)*, out of which seven were of farmers and five were agricultural labourers.

Sangatpura with a population of 3,099, has had 32 suicides (*including 4 cases of multiple suicides*). This village has 507 households with 1,140 workers, out of which 10.5 are main workers. The gender ratio of Sangatpura is low: 890 and 0.76 in the 0–6 age group. Literacy rate here is also low: 45.11 per cent. Out of the total suicides, 18 have been farmers and four were agricultural labourers. Sangatpura has had double suicides. In the family of Chet Singh, one son Manjeet, committed suicide in 1998 and another son, Jagdev Singh, committed suicide in November, 2003.

Sekhuwas is a medium size village with a population of 2,459 with 425 households and only 974 total workers, out of which 765 are main workers and 296 are marginal workers. Also, its gender ratio of 665 is among the lowest in the age group of 0–6 years and the literacy rate is a low at 36.2 per cent. Eight persons have committed suicide; all were farmers.

Shadihari is a big village. Its population is 4,387 with 739 households and 1,927 total workers. It has a literacy rate of 42.8 per cent. Its sex ratio amongst elders and 0–6 age group is almost the same with 84.9 and 85.5 respectively. Nineteen persons have committed suicide. All of them owed ₹2 lakh or more to moneylenders and 10 of them have left behind families with little children. Eleven were farmers and eight were agricultural labourers.

Thaska is a very small village with a population of 730 and 130 households. Its working population is small: only 69 main workers and 149 marginal workers. One can see that the burden of those who are working is quite heavy. Its gender ratio is 968 and its literacy rate is 53.7 per cent. Sixteen persons have committed suicide (*including 1 case of multiple suicides*), out of which four were of farmers and 11 were agricultural labourers. The village has about 14 households of farmers while the remaining 116 are agricultural labourers and people in occupation related to agriculture.

Four Additional Villages

For the purpose of this study four more village haven been chosen. These are under the Lehra police jurisdiction, although they actually fall within the Sunam *tehsil*. These are Fatehgarh, Gandhuan, Gobindgarh Jaijian and Phuleda.

Fatehgarh has a population of 2,180 with 369 households and 742 of 2,180 with 369 and 742 total workers and 1,438 non-workers. It has a literacy rate of 4.4 per cent. Eleven persons have committed suicide (*including 1 case of multiple suicides*); all were aged between 35 and 45 years. Five of the victims were farmers, five were agricultural labourers and one was a shopkeeper.

Gandhuan is a big village with a population of 4,550 and 765 households. The village has 1,540 workers, out of which 1,445 are main workers, 96 marginal workers and 3,010 non-workers. Its literacy rate is 42.8 per cent. Two persons had committed suicide: an agricultural labourer and a woman.

Gobindgarh Jaijian is a medium-sized village of 2,783 people and 447 households. There are 1,358 workers, out of which 1,258 are main workers and 100 are marginal workers. The literacy rate is 48.1 per cent. Twenty-five persons have committed suicide, out of which 11 were farmers and 14 were labourers. Suicides of four women have been recorded; they were aged between 25 and 32; two were unmarried.

Phuleda is again a medium-sized village with 2,128 residents and 368 households. It has only 846 workers, out of which only 670 are main workers. The literacy rate is 41.7 per cent. Thirty-one persons have committed suicide in Phuleda (*including 2 cases of multiple suicides*); of these 24 were farmers and only seven were agricultural labourers, out of which two were women.

Andana Block

Forty villages fall within Andana Block and of these, 18 villages come under strong Haryanvi influence and follow Haryanvi customs. Also, 25 villages bear the brunt of the Ghaggar's annual flooding in the monsoon season.

Andana is a big village with a population of 4,586 and 727 households. Andana has 2,328 total workers, out of which 2,169 are main workers and 159 marginal workers. The village has a literacy rate of 52.3 per cent and a low gender ratio of 870. This village is predominantly Haryanvi in nature and highly flood-prone, although the village itself is located at a height so that the houses are saved even when the fields are submerged. Thirty-five persons have committed suicide (*including 3 cases of multiple suicides*), out of which three were women. Twenty-five farmers committed suicide and 10 were agricultural labourers. The occupation of one victim could not be verified. One also finds cases of multiple suicides in one family. Ram Swaroop lost two sons in the span of three years. His elder son Mohinder Singh (25) committed suicide in July, 2000, and about three years later, the younger son Suresh committed suicide. Interestingly at the time of Mohinder Singh's suicide, the family had 3 acres of land and a debt of ₹150,000 but by 2004, and in spite of having sold all the land and becoming an agricultural labourer, the debt had risen to ₹2 lakh. This is not uncommon: persons trapped in debt see their liability grow heavier with every passing year as the *arthiya* charges compound interest at the rate of ₹4 to ₹6 for every hundred rupees of the debt, and this amount is compounded twice a year.

Badalgarh is a small village of 1,729 people with 316 households and 870 total workers. Badalgarh's gender ratio is low at just 745, but the gender ratio in the age group of 0–6 is the lowest in the region—a mere 471. The literacy rate is 39.4 per cent. Seven persons have committed suicide: all were farmers, including a woman. Two of the victims owned 12 acres of land or more. People of Badalgarh follow Punjabi customs and the village and is not flood prone.

Bahmniwala is a small village of 168 households and 1,054 population. The customs here are predominantly Haryanvi. The village is not flood-prone. It has a total working population of 601 with all these as main workers. The literacy rate is 56 per cent.

Thirteen persons committed suicide out of which four were farmers and five were agricultural labourers. The occupation of four victims could not be verified. It has been noted that in almost all the cases, the agriculturists sold land to clear debts. This has resulted in rendering nearly the entire village landless now.

Banga is a village where people follow Haryanvi customs. This is a relatively small village with a population of 2,585 residents and 1,111 households. It has 1,159 workers, out of which 778 are main workers and 381 are marginal workers. The literacy rate is 40.5 per cent. Thirty-three persons have committed suicide (*including 3 cases of multiple suicides*), out of which 25 were farmers and 8 were agricultural labourers. One woman in Banga committed suicide.

Baopur is a 'Haryanvi' village and flood-prone; its population is 1,524 with 251 households. Main workers number only 490. Baopur has a gender ratio of 917 which has declined over the years; in the age group of 0–6 it is 861. The literacy rate is 48.4 per cent.

Fifteen persons have committed suicide (*including 2 cases of multiple suicides*); nine farmers and six agricultural labourers, including one woman.

Benarsi is a 'Haryanvi' village and is flood-prone. It stands on a low mound, so flood water does not enter the houses. It has a population of 2,585 people with 375 households. Benarsi has 1,169 total workers and a literacy rate of 51.2 per cent. Fifteen persons have committed suicide (*including 1 case of multiple suicides*), out of which 10 were farmers and 5 were agricultural labourers. Man Paul and Shamsher Singh each owned 3 acres when they committed suicide in 2000; since then their families have had to sell the land and now survive as labourers. Some of the debt is still pending.

Bhoolan has a population of 3,725 people and 531 households with 1,705 total workers and 44.7 per cent literacy rate, according to the census Abstract of Punjab 2001. On personal verification however, it was noted that the village has 600 households and a total of 2,000 acres in the village. This means that holdings per household are very small and making debt in evitable; 57 persons have committed suicide (*including 6 cases of multiple suicides*). The years in which the maximum number of persons committed suicide were: 1990, 1992, 1994, 1996, 1998 and 2004. Victims included 47 farmers, nine agricultural labourers and one whose occupation could not be verified. On 2 March 1998, Fota Singh, son of Ramdas Singh, committed suicide by jumping into the Bhakra and his body was never found. On 15 June 2004, Mihan Singh sold all his land and later committed suicide. Unable to bear the shock, his wife soon died. Two mentally-challenged sons survive. Village charity keeps them alive but their existence is wretched.

Bhunder Bhaini is a flood-prone village where Punjabi customs prevail. Residents number just 1,051 and there are 170 households. Workers number 331. Literacy rate is 42.6 per cent. There have been five suicides, of which three were farmers and two were agricultural labourers.

Bishenpura Khokhar has a population of 1,280 with 215 households. Total workforce is 640 and the literacy rate is 46.5 per cent. Its gender ratio shows a drastic downward trend with elders having 850 and the gender ratio between the age 0–6 being only 504. It is a 'Punjabi' village and not prone to flood. Thirteen persons, 10 agricultural labourers and the rest agricultural labourers, have committed suicide in this village.

Bushera has a population of 2,666 with 403 households. Its total workers are 1,374, out of which 749 are main workers and 540 are marginal workers. The literacy rate is 40.1 per cent. Bushera is a flood-prone village. This village follows Haryanvi customs.

Twenty-five persons have committed suicide (*including 2 cases of multiple suicides*): seven farmers, 15 agricultural labourers and three in other occupations. Jarnail Singh committed suicide in 1986 and his father Juga Singh committed suicide in 1995.

Chatha Gobindpura has 2,555 residents with 366 households and 678 workers. In other words, only about a quarter are working. The literacy rate is only 42.5 per cent. Thirteen persons committed suicide out of which 6 were in 2002 alone. Seven of these were farmers and six were agricultural labourers.

Chandu is a flood-prone village with Haryanvi customs. Residents number 1,289 with 202 households, and there are 438 workers. Its gender ratio in the age group 0–6 is 1,024, which is quite good. Four agricultural labourers of this village have committed suicide.

Dudian has a population of 2,944 with 484 households and a literacy rate of 34 per cent. It has 742 total workers with 655 main workers which is even less than one-fourth of its population. This is a 'Punjabi' non flood-prone village. Twenty-two persons have committed suicide (*including 1 case of multiple suicides*), of which 14 were farmers, seven were agricultural labourers and one victim's occupation could not be verified. Ajmer Singh committed suicide on 18 July 1994; at that time he had 2 acres and ₹2 lakh debt. By the time his son Pala Singh committed suicide on 11 August 2001, the family was down to 1 acre and the debt stood at ₹15 lakh.

Ganauta is a small village with only 127 households and 929 people. It has only 296 total workers, out of which six are marginal workers. Surprisingly, the literacy rate here is a high at 60.9 per cent. It is a 'Punjabi' village and is flood prone. Fifteen persons have committed suicide (*including 1 case of multiple suicides*), out of which seven were agricultural labourers.

Ghamur Ghat has a population of 1,382 with 215 households and 422 total workers. The literacy rate is 51.6 per cent. It is culturally 'Punjabi' and safe from floods. Five persons have committed suicide out of which three were agricultural labourers.

Gulahri is a medium-sized village with 3,038 residents and 456 households. The literacy rate is 50.9 per cent. It has 1,560 total workers, out of which 1,293 are main workers. Twenty-one persons have committed suicide (*including 2 cases of multiple suicides*), out of which three were farmers, and the occupations of two persons could not be verified. The remaining 16 victims were agricultural labouers. Ram Swaroop committed suicide in 1989 and his son Subhash committed suicide on 10 April 2002. Ram Swaroop (aged 58 at the time of death) had been a *sarpanch*.

Handa is a 'Punjabi' village and safe from flood. Its residents number at 1,010 and it has 141 households. The total workers number 568 and the literacy rate is 47.8 per cent. Seven persons have committed suicide, out of which three were farmers.

Hamirgarh is a 'Punjabi' village and is flood-prone. It is a medium-sized village with 2,470 population and 448 households. It has 953 total workers with 762 main and 191 marginal workers. Its literacy rate is 46.1 per cent. Twenty-nine 29 persons have committed suicide (*including one case of multiple suicides*), out of which 18 were farmers and 11 were agricultural labourers. One was a woman.

Harigarh Gehla is a 'Punjabi' village and safe from floods. Since 1990, 23 persons have committed suicide. This is a high number since the population is only 941 (170 households). It has a total of 482 workers with a 52.6 per cent literacy rate. Twenty farmers have committed suicide (*including 3 cases of multiple suicides*), of which 16 were agricultural labourers. Among the victims are two women. In 2005, Jasmat Singh (age 38), son of Dayal Singh, had to sell his small holding to repay his debt; in June that year he drank pesticide but was saved. He is still under huge debt and the loss of his land plus his inability to perform the marriage of his Daughter has left him deeply depressed. He regards earning as a labourer as a shame on himself and his family.

Hauti Khurd is a *bechirag* [deserted] village in which no one lives any longer.

Hotipur is a flood-prone village with a population of 944 and 164 households following Punjabi customs. It has a total of 287 workers and a literacy rate of 65.4 per cent. Seven persons have committed suicide. It has the highest literacy rate in the subdivision.

Kadial is a 'Punjabi' village and prone to floods. Its residents number at 2,456 and there are 348 households. It has 797 total workers, out of which 549 are main workers and 248 are marginal workers. Its literacy rate is 55.7 per cent. It has had three suicides, out of which two were of agricultural labourers.

Karoda is a 'Haryanvi' village with 303 households and 1,893 residents, but only 555 total workers and a literacy rate of 45.8 per cent. Twenty-seven persons have committed suicide (*including 3 cases of multiple suicides*), out of which 20 were farmers. Brothers Amrit and Jai Singh both committed suicide in 1994. Dufa Ram committed suicide in 1995 and his brother Rameshwar committed suicide on 5 September 2001. At the time of Dufa Ram's death, the family had 3 acres but by 2001, when Rameshwar committed suicide, they were down to just two kanal. Chander Bhan committed suicide in July,

2001 and his son Gurnam committed suicide on 4 January 2004. Another Gurnarn, who committed suicide 23 June 2004, was a cousin of Dufa Ram. These families had holding in excess of 20 acres but had to sell off and now each family is left with just 1 acre.

Khanauri Kalan and Khanauri Khurd for the purposes of census, have been clubbed together and some part of them have been included in Khanauri town. Khanauri Kalan has a population of 756 people with 132 households and 53.1 per cent literacy rate. They have a total of 390 workers. Thirteen 13 persons have committed suicide (*including 2 cases of multiple suicides*), out of which eight were farmers and four were agricultural labourers. The occupation of one victim was not known. A husband and wife, Jasbir Kaur and Basant Singh, committed suicide on 16 July 2004. In Khanauri Khurd, six persons have committed suicide and all agricultural labourers.

Maha Singhwala is a 'Punjabi' village with a population of 125 and 213 households. It has a total of 660 workers out of which 57 are marginal workers. The literacy rate is 46.2 per cent. Seventeen persons have committed suicide out of which seven were farmers.

Kudni has a population of 1,041 with 151 households and 495 workers, of whom 235 are main workers and 268 are marginal workers. The gender ratio in both the adult and 0–6 years age groups is 817. The literacy rate is 55.1 per cent. Two agricultural labourers of this village committed suicide in 2001.

Mandvi is a flood prone Haryanvi village. It is a big village: population numbers at 5,407 with 932 households. It has a total of 1,725 workers out of which 1,516 are main workers and 209 are marginal workers. The literacy rate is 54.3 percent. Seven persons have committed suicide out of which four were agricultural labourers.

Makror is a 'Haryanvi' village with a population of 2,975 and 429 households. It has 1,390 total workers out of which 963 are main workers and 427 are marginal workers. The literacy rate is 49.6 per cent. Thirteen persons have committed suicide, out of which six were farmers and seven were agricultural labourers. One of the victims was a woman.

Maniana is a village of 3,247 people and 542 households, with a total of 1,570 workers. Haryanvi customs prevail here. The literacy rate is 40.1 per cent. It has had 15 suicides (*including 1 case of multiple suicides*); eight farmers and seven farm labourers. Resham Singh had sold 5 acres out of his 10-acre farm on 3 May 2001, but could not bear this and committed suicide 15 days later on 30 May 2001.

Moonak for census purposes, this village has been added to Moonak Township. One finds both Punjabi and Haryanvi customs equally prevalent here. The village is flood prone. Thirty-four persons have committed suicide (*including 2 cases of multiple suicides*), of which 27 were farmers and six were agricultural labourers. The occupation of one victim was unknown. Three women committed suicide. A husband and wife, Palpi and Ram Kumar, committed suicide in 1998. In 1999, Satinder Singh and his son attempted suicide. Satinder died but his son was saved.

Navagaon is a 'Punjabi' village that is prone to floods. It is medium sized with a population of 2,653 and 478 households. The workforce numbers at 1,313 and the literacy rate is 43.6 per cent. Twelve farmers have committed suicide, out of which two victims had incurred very high debts—₹6 lakh and ₹10 lakh respectively.

Phoolad is a flood-prone village of 1,504 residents and 251 house holds that follows Punjabi customs. It has 647 workers and a literacy rate of 50.7 per cent. Fourteen persons have committed suicide (*including 2 cases of multiple suicides*).

Rajalheri is a small 'Punjabi' village of 1,034 persons and 170 households. It has 470 workers and the literacy rate is 46 per cent. Half the village land is flood prone. One farmer, indebted to the tune of ₹7 lakh, committed suicide.

Ramgarh Gujran is a small 'Haryanvi' village safe from floods; its population is 898 with 153 households. Workers number at 499 and the literacy rate is 52.2 per cent. Seven persons have committed suicide; four were farmers and two were agricultural labourers. The occupation of one victim could not be verified.

Rampura is a flood-prone village that follows Punjabi customs. Population stands at 2,572 with 414 households. Main workers number at 969 and 27 are marginal workers. The literacy rate is 51.9 per cent. Twelve persons have committed suicide (*including 2 cases of multiple suicides*), of whom seven were agricultural labourers.

Salemgarh is a 'Punjabi' village and is highly flood prone. It has a population of 2,246 with 406 households. It has 1,787 total workers with 48.4 per cent literacy rate. Fifteen persons have committed suicide (*including two cases of multiple suicides*), of whom 12 were farmers and three were agricultural labourers.

Suratpur Nabhani is a *bechirag* village.

Shahpur Theri is a small flood-prone Punjabi village. Its population is 970 with 155 households. It has 211 total workers and 56.3 per cent literacy rate. Seven persons have committed suicide out of which five were farmers, including a woman, and two were agricultural labourers.

Shergarh is a Punjabi village which is also called Sinha Singh Wala. It has a population of 1,301 with 230 households and 711 total workers. Fifteen persons have committed suicide. It has had nine suicides by farmers and five by farm labourers. The occupation of one victim could not be verified.

Surjan Bhaini is a flood-prone Punjabi village. It is very small with only 40 households and a population of 277. It has 82 total workers and a literacy rate of 53.7 per cent. Three suicides took place here, among these, one was an agricultural labourer.

Index

About the Authors

Aman Sidhu lost her life in a tragic road accident in 2006. This book is based on her PhD research, focusing on rural suicides in Punjab. Aman Sidhu was 34 years of age when this research began. She had recently returned from the United Kingdom having obtained an MBA degree, prior to which she had completed her Master's in Sociology from Panjab University. Aman's interest in rural suicides developed while working with the Movement Against State Repression (MASR), an NGO, for which she was conducting research in the Lehra and Andana Blocks of Moonak subdivision in Punjab's Sangrur district. She decided to take this work forward, by enrolling for a PhD based on her field-work. Aman was also instrumental in starting the JSJ Degree College at Gurney Kalan, in addition to various vocational centres for village girls.

As her PhD research had led to the compilation of a significant amount of statistical data and interviews, her father, Mr Inderjit Singh Jaijee, decided to take the work forward as a memorial to her.

Inderjit Singh Jaijee has been a long serving advocate of people's rights and dignity in the Punjab region. In 1985 he was elected to the Punjab Legislative Assembly; he has served as President of the Indian Minority and Dalit Front; has worked for more than 15 years with the Movement Against State Repression (MASR); and is a founding member of the Rescue and Revival Mission of the Baba Nanak Educational Society.